Physico-chemical Properties of Nucleic Acids

Physico-chemical Properties of Nucleic Acids

edited by J. DUCHESNE
Department of Molecular
and Atomic Physics
University of Liège
Belgium

VOLUME 3: *Intra- and Intermolecular Interactions,*
Radiation Effects in DNA Cells,
and Repair Mechanisms

ACADEMIC PRESS
London · New York
1973

A subsidiary of Harcourt Brace Jovanovich, Publishers

ACADEMIC PRESS INC. (LONDON) LTD.
24/28 Oval Road
London NW1

United States Edition published by
ACADEMIC PRESS INC.
111 Fifth Avenue
New York, New York 10003

Library of Congress Catalog Card Number: 72–84360
ISBN: 0-12-222903-7

PRINTED IN GREAT BRITAIN BY
Adlard & Son Ltd., Bartholomew Press, Dorking

Contributors to Volume 3

F. W. Benz, *National Institute for Medical Research, Mill Hill, London N.W.7, Great Britain.*

M. Dourlent, *Centre de Biophysique Moléculaire, Orléans, France.*

M. Errera, *Laboratoire de Biophysique et Radiobiologie. Faculté des Sciences, Université libre de Bruxelles, Belgium.*

C. Hélène, *Centre de Biophysique Moléculaire, Orléans, France.*

M. Leng, *Centre de Biophysique Moléculaire, Orléans, France.*

L. S. Lerman, *Department of Molecular Biology, Vanderbilt University, Nashville, Tennessee, U.S.A.*

M. G. Ormerod, *Chester Beatty Research Institute, Institute of Cancer Research, Royal Cancer Hospital, Clifton Avenue, Belmont, Sutton, Surrey, England.*

M. Radman, *Laboratoire de Biophysique et Radiobiologie, Faculté des Sciences, Université libre de Bruxelles, Belgium.*

H. Richard, *Centre de Recherches Paul Pascal Domaine Universitaire, France.*

G. C. K. Roberts, *National Institute for Medical Research, Mill Hill, London N.W.7, Great Britain.*

J. Rommelaere, *Laboratoire de Biophysique et Radiobiologie, Faculté des Sciences, Université libre de Bruxelles, Belgium.*

Preface

Progress in the application of physics and chemistry to biology has been particularly important in a number of areas, and this book brings together recent work in these fields.

Nearly all the topics have been approached in the spirit of molecular biology and biophysics, where recent advances are far from having exhausted their impact.

Though the book has been subdivided into three parts for practical reasons, the unity of the whole work has been largely preserved and each volume is also an entity as it stands.

The first volume is mainly concerned with the intrinsic properties of nucleic acids, considered as macromolecules, and their components, and this study is completed by the analysis of different types of bindings or interaction mechanisms, including photodynamic and radiation effects, as well as fluorescence. In the second volume, the emphasis is put on structural studies and especially on conformational changes, using spectroscopic techniques as well as methods of thermodynamics and hydrodynamics. The stage of specific biological functions is attained in the last volume, with some considerations on repair mechanisms in relation to the general problem of evolution.

"The Physico-chemical Properties of Nucleic Acids" is intended to provide thought provoking material for research scientists, whether they are biologists, chemists or physicists. At the same time, it should be a source of information and reference for graduate students in these fields.

It is particularly hoped that these papers may help to stimulate the search for a better understanding of the correlation between structure and function. This understanding can only be founded on a detailed knowledge of the molecular properties of the basic substances.

Lastly, it is a great pleasure for the Editor to acknowledge the excellent cooperation of the publisher, who has been helpful in every way.

November 1972 JULES DUCHESNE

Foreword

The chemist, if he encounters a new substance, is accustomed to look first for its macroscopic properties in order to correlate them with underlying structural features. Such properties instantaneously suggest to him some picture of the molecule, and often it is only the material quality which stimulated his interest in particular substances.

If we think of nucleic acids, such a correlation does not usually come to our mind. In fact, most of us working on one or the other aspect of this fascinating molecule, have not even seen any pure crystalline material, not to speak of particular samples such as isolated gene material or single uniform *t*RNA batches, which actually are the objects of the most exciting studies in molecular biology.

Speaking of nucleic acids we usually do not associate with them any characteristic material property; we rather think of some abstract quality: information, instruction, translation, etc. We see before us sequences of letters, such as

. . . pApUpGpCpGpUpApUpApGpCpApUp . . .

and we think of a message. Of course, this is an exaggeration. Most of us—encountering any analysed sequence—would immediately start to play around with it and fold it up, e.g. as

$$\begin{array}{r} \ldots \mathrm{pApUpGpC}^{\mathrm{pG}}\mathrm{pU} \\ \mathrm{pA} \\ \ldots \mathrm{pUpApCpG}_{\mathrm{pA}}\mathrm{pU} \end{array}$$

This shows that we actually associate with each letter also a particular physical interaction unique to this species of macromolecules.

It is this interaction, this exclusive way of complementary pairing of bases which is behind the abstract property of "code reading" or "information transfer", and it was the realisation of this quality which led Watson and Crick to their epochal discovery (quantitatively manifested in Wilkin's X-ray diffraction data).

How far our interpretations are guided by abstract reasoning became especially apparent in the deciphering of the genetic code. With 20 amino acids (plus some punctuation symbols) to be coded by the four bases (i.e. A, U (or T), G and C) it was "obvious" that the code had to be a triplet code—as indeed has been confirmed by the work of Nirenberg, Matthaei, Khorana and Ochoa. However, this was "obvious" only by logical arguments: a doublet code could only provide 16, i.e. less than 20 codons, a quadruplet code would be uneconomical in providing too many, i.e. 256 codon units. But how could molecules be so intelligent as to accept a logical argument? The answer is: It is the physical behaviour of the nucleotide sequences which determines the optimal choice among the different possible associations and the one chosen appears to be in agreement with our rational logic, because it offers evolutionary advantages with respect to precision and speed of information transfer. Precision requires interactions of sufficient stability, thus it involves "stickiness" which limits the rate of information transfer. The codon–anticodon interaction therefore must be optimised to involve:

1. sufficient functional capacities (i.e. requiring more than twenty combinations)
2. distinctive recognition (requiring at least base triplet interactions) and
3. sufficiently low stickiness (keeping the codon-unit as small as possible).

What I wanted to say is that all abstract qualities of nucleic acids which we associate with their function to store, transfer, and process information are reflected by certain unique physico-chemical properties. It is not sufficient to have just macromolecular species resembling a sequence of different digits. The "digits" in addition must provide very specific physical interactions to cause the inherent property of self-instruction and code formation. This quality, unique to the nucleic acids (and their interactions with proteins) provides the capacity of self-organisation according to—or in agreement with—our abstract principles of purposefulness, usefulness and rationality.

In this situation it is highly desirable to use any available experimental tool to enhance our knowledge about the "Physico-chemical Properties of Nucleic Acids". The three volumes which appear under this title offer a large repertoire of studies. Not all of them may be equally relevant for an understanding of the characteristics of information processing, which also involves highly specific interactions of nucleic acids with proteins. Nevertheless, all these studies will finally contribute to our basic understanding of those properties, which are behind the structural features, specific interactions, and dynamic performances of these unique macromolecules. In our age of molecular biology it may seem to be somewhat fashionable to do research in the field of nucleic acids. However, there will be a long persisting interest in this field before our knowledge about the molecular details will have brought about a complete understanding of the sophisticated organisation of the genome of a highly developed cell.

Spiegelman once characterised the central role of nucleic acids by saying jokingly: "The evolution of life is a trick of nature to ensure a faster and better reproduction of the nucleic acids".

November 1972 MANFRED EIGEN

Contents Volume 3

Intra- and Intermolecular Interactions, Radiation Effects in DNA Cells, and Repair Mechanisms

Contents Volume 1

"Electrical, Optical and Magnetic Properties of Nucleic Acids and Components"

Contents of Volume 2

"Structural Studies on Nucleic Acids and Other Biopolymers"

CHAPTER 17

Renaturation of DNA in the Absence and in the Presence of Cu^{2+} Ions

H. RICHARD

Centre de Recherches Paul Pascal, Domaine Universitaire

I. Introduction

The fact that DNA and certain polynucleotides exhibit a Watson and Crick double helix structure is now well known. This ordered structure is the native one in which the complementary base pairs Adenine-Thymine (AT) and Guanine-Cytosine (GC) are parallel to one another and perpendicular to the helix axis. Denaturation is the physical treatment which destroys this order. Renaturation is the reverse process, that is, the transformation from the disordered state to the ordered one.

II. Renaturation Kinetics of Polynucleotides in the Absence of Cupric Ions

As early as 1963, Marmur and his co-workers (Marmur *et al.*, 1963) completed a bibliographical study of renaturation of polynucleotides in the absence of ions such as Cu^{2+}. Wetmur and Davidson (1968) clearly summed up the known results to that time. However there has been little subsequent advance of our knowledge in this field since this date. At the risk of repeating previously reported results we include here, for the convenience of the reader, all known results in a slightly modified form.

A. MATERIALS AND METHODS

From Table I it can be seen that various types of polynucleotides have been studied and that the most frequently used technique to measure the degree of renaturation α of the DNA as a function of time is u.v. spectrophotometry (hypochromicity). If α is obtained by measuring a property P, the optical density for instance, it is defined by the relation:

$$\alpha_P = \frac{P_1 - P}{P_1 - P_2}$$

where P is the value of the property P at time t and P_1 and P_2 are its value at the initial and final times respectively.

TABLE I

Authors	Polynucleotides	Techniques
P. D. Ross and J. M. Sturtevant (1960).	Poly (A+U)	u.v. spectrophotometry (Stopped-Flow)
J. Marmur and P. Doty (1961).	Bacterial DNA	u.v. spectrophotometry Transforming activity
L. F. Cavalieri, Th. Small and N. Sarkar (1962).	Bacterial DNA	u.v. spectrophotometry
R. B. Inman and R. L. Baldwin (1964).	Poly dI : $\overline{\text{dBC}}$	u.v. spectrophotometry
K. J. Thrower and A. R. Peacoke (1966).	Bacterial, Phage DNA	u.v. spectrophotometry
J. A. Subirana and P. Doty (1966).	Bacterial DNA	u.v. spectrophotometry
R. D. Blake and J. R. Fresco (1966).	Poly (A+U)	u.v. spectrophotometry
J. G. Wetmur and N. Davidson (1968).	Bacterial, Phage DNA	u.v. spectrophotometry
R. J. Britten and D. E. Kohne (1968).	Bacterial, Vertebrate DNA	Hydroxyapatite column
F. W. Studier (1969).	Phage DNA	Sedimentation velocity
J. G. Wetmur (1971).	Phage DNA	u.v. spectrophotometry

B. VARIABLES

The alteration of the value of some of the variables has the consequence of changing the rate of renaturation v.

A first attempt at enumeration of the variables has been done by Marmur and Doty (1961) with the purpose of defining the optimal conditions of the renaturation of DNA.

1. *Source of the DNA*

DNA samples from different sources can be distinguished by both base composition and by their complexity. On one hand the speed of renaturation v increases as a function of the percentage of GC base pairs in the DNA (Wetmur and Davidson, 1968), while on the other, all other things being equal, the speed of renaturation v of bacteriophage DNA is greater than that of bacterial DNA which in turn is greater than that of mammalian DNA. Britten and Kohne (1968) quantitatively established that the rate of renaturation decreased as a function of the length of the genome. In the same year Wetmur and Davidson (1968) reported an empirical relation between the rate constant k_2, the length of single strands of DNA, and the number of base pairs in non-repetitive sequences N:

$$k_2 = 3.10^5 . L^{0 \cdot 5}/N$$

2. *Renaturation Temperature* T_2†

There are two distinguishable cases: that of synthetic polymers and that of DNA.

The formation of poly (A+U) (Ross and Sturtevant, 1960) and poly dI: $\overline{\text{dBC}}$ (Inman and Baldwin, 1964) is such that the speed of transformation increases as the temperature is reduced. If T_m is the melting temperature, the rate constant k is a positive linear function of $(T_m - T_2)$ (Ross and Sturtevant, 1960).

The renaturation rate of DNA passes through a maximum as the temperature decreases (Wetmur and Davidson, 1968; Marmur and Doty, 1961); the maximum occurring at around 25° below T_m, depending on the GC composition of the DNA.

Studier (1969) has shown that the degree of folding of single strands increases with temperature and that the renaturation rate of folded strands is less than that of the unfolded forms. In fact the effect of temperature on the rate v could be the result of the temperature effect on the folding of the single strands.

3. *Ionic Strength* μ *of the Solvent*

The degree of renaturation is optimal when the ionic strength, obtained by the addition of sodium ions, is greater than 0·4 M ($\mu > 0 \cdot 4$ M). All the investigators having studied the influence of ionic strength on the rate of renaturation (Wetmur and Davidson, 1968; Ross and Sturtevant, 1960; Inman and Baldwin, 1964; Thrower and Peacoke, 1966) report concordant results affirming that v increases with μ. Only Studier (1969) claims that v passes

† T_1 is used to denote the denaturation temperature.

through a maximum with increasing μ. This would seem to be connected to the combined effect of μ on the speed of transformation and the folding of the single strands.

4. *Solvent Viscosity* η

If the viscosity of the solvent is increased by the addition of sucrose (Subirana and Doty, 1966; Thrower and Peacoke, 1966; Wetmur and Davidson, 1968), glycerol or $NaClO_4$ the rate of renaturation decreases. In contrast, increasing η by the addition of Na polyacrylate or native T4 phage DNA results in an increase in v.

5. *pH of the Solvent*

The rate of renaturation increases with the concentration of DNA (Inman and Baldwin, 1964; Subirana and Doty, 1966).

6. *Molecular Weight–Length*

The degree of renaturation increases with molecular weight (Marmur and Doty, 1961).

Consider the equation previously cited (Wetmur and Davidson, 1968):

$$k_2 = 3.10^5.L^{0.5}/N$$

7. *Degree of Folding of Single Stranded DNA*

Having prepared T7 phage single stranded DNA, Studier estimated the degree of folding by simultaneously measuring the optical density and the sedimentation velocity. It is possible to alter the folding of single strands by modifying the temperature, the ionic strength or the pH. Studier established that the rate of renaturation v is reduced as the degree of folding increases (Studier, 1969).

8. *Theoretical Studies*

The various theoretical treatments account for, in a satisfactory manner, the temperature effect and the length of the DNA (Ross and Sturtevant, 1960, 1962; Saunders and Ross, 1960; Wetmur and Davidson, 1968; Wetmur, 1971).

a. *Effect of temperature.* The simplest model which accounts for the variation of the rate of formation of poly (A+U) (Ross and Sturtevant, 1960) as a function of temperature is that of Saunders and Ross (1960). They consider that processes in polymers consist of a succession of identical steps carried out by each one of the monomer. Each step is supposed to be reversible and as a result there corresponds to each step two rate constants: one being k_f for the reaction under consideration while the other k_b is for the reverse reaction. Only one step is different from the others, that being the initiation step for which the rate constant is k_i. The steady state approximation is assumed to be applicable. If the reaction is of the type:

$$A+B \rightarrow C_1 \rightarrow C_2 \ldots \rightarrow C_n$$

the rate of the reaction can be expressed by:

$$v = k_i\,[A]\,[B]\left(1-\frac{k_b}{kf}\right)$$

Furthermore if it is assumed that the rate constants can be described by the Arrhenius equation, the expression for v becomes:

$$v = C\exp(-\Delta H^*/RT)\,[1-D\exp(-\Delta H/RT)]$$

where C and D are constants, ΔH^* the activation energy and ΔH the enthalpy of formation of a step. At the melting temperature T_m, $v=0$ by definition and thus $[D\exp(-\Delta H/RT)]=1$. This latter expression is less than one when the reaction takes place. By adjusting the values of ΔH^* and ΔH the experimental results of Ross and Sturtevant (1960) for the formation of poly (A+U) can be accounted for.

The preceding model has been discussed and modified by Flory (1961), Ross and Sturtevant (1962), Kallenbach, *et al.* (1963), but the mode of calculation has remained unchanged. Finally Wetmur and Davidson (1968) introduced into this model the influence of random base sequences for DNA and in this manner explained the curve $v=f(T)$ observed for DNA (Marmur and Doty, 1961; Wetmur and Davidson, 1968).

b. *Effect of Chain Length.* The variation of the rate of renaturation as a function of chain length can be explained by an excluded volume theory. To do this, Wetmur (1971) has calculated the probability of two random coils approaching one another in such a manner that their centres of gravity are at a distance d and subsequently determined the rate of renaturation of the random coils situated at this distance. The rate of renaturation depends on the relative probability $P(d)$ of a configuration and also on the probability of overlapping of a given pair of segments.

This theory accounts very well for the second-order rate constant k_2 varying as $L^{0.5}$.

C. ANALYSIS OF THE KINETIC RESULTS

In order to process the kinetic curves obtained under different experimental conditions, different authors have tried to categorise the transformation studied. As is shown in Table II the order of the reaction is often poorly defined. It appears that the most recent studies (Britten and Kohne, 1968; Wetmur and Davidson, 1968) favour an assignment of second order.

TABLE II

Authors	Reaction order n
P. D. Ross and J. M. Sturtevant (1960).	$n=2$ at the beginning of the reaction $\rightarrow n=1$ near the end; the more μ diminishes, the more $n \rightarrow 1$.
L. F. Cavalieri, Th. Small and N. Sarkar (1962).	$n=1$ at $T=60°C$; second-order effect becoming more important when $T=70–80°C$
R. B. Inman and R. L. Baldwin (1964).	$n=2$.
K. J. Thrower and A. R. Peacoke (1966).	$n=2$ (first 3 hours) $\rightarrow n>2$ (longer times).
J. A. Subirana and P. Doty (1966).	$n=2$ with deviations; as increasing μ the deviations, relative to a second-order curve, increase.
R. D. Blake and J. R. Fresco (1966).	$n=2$ under the conditions where only poly (A + U) is formed.
J. G. Wetmur and N. Davidson (1968).	$n=2$ results reported just to 75% of the complete transformation.
R. J. Britten and D. E. Kohne (1968).	$n=2$.

D. RENATURATION MECHANISMS OF DNA AND SYNTHETIC POLYNUCLEOTIDES

The mechanism most frequently suggested for the renaturation of DNA as well as the synthetic polynucleotides is a two step transformation: nucleation followed by a zippering-up (Flory, 1961; Kallenbach *et al.*, 1963; Marmur and Doty, 1961; Ross and Sturtevant, 1960, 1962; Studier, 1969; Subirana and Doty, 1966; Wetmur and Davidson, 1968).

Nucleation is the formation of a minimum number of complementary base pairs permitting the strands to remain face to face. The fact that the total process seems to be second order suggests that this first step, a bimolecular process, is slow. The zippering-up, a unimolecular process, would in contrast be very rapid.

A certain number of complementary hypotheses have been made in order to explain deviations from second-order kinetics. Cavalieri and collaborators in studying DNA (Cavalieri *et al.*, 1962) imagined a mechanism involving two sorts of molecules: one being a single strand S and the other with strands partially separated (NS). In contrast, Thrower and Peacoke (1966) who also studied DNA, explained previous results using purely experimental considerations. Ross and Sturtevant (1960) put forward three possible hypotheses to make the deviation from second-order kinetics observed during the formation of poly (A + U) in solution of low ionic strength comprehensible:

The rate of cooperative ordering is much slower than that of nucleation.
The base pairs are randomly bound during nucleation.

The rate of complex formation depends on the ionic strength of the solvent as this affects the conformation of the polymers.

However Blake and Fresco (1966) later showed that the alteration of rate from second to first order is simply due to simultaneous formation of poly (A+U) and poly (A+U+U).

The third hypothesis of Ross and Sturtevant (1960) has been verified in the case of phage DNA by Studier (1969).

III. DNA Renaturation Kinetics in the Presence of Cupric Ions

This particular aspect of DNA renaturation has only been studied kinetically by Richard and Pacault (1968, 1969, 1970, 1971). The system, in thermodynamic terms, has the following constraints: the pressure P, the temperature T and the number of mol of constituents n_i. The constituents are:

DNA, extracted from different sources,
the salt determining the ionic strength of the solvent,
a certain quantity of Cu^{2+} ions, determined by the relation:

$$a = Cu^{2+}/DNA-P.$$

A. MATERIALS AND METHODS

The renaturation kinetics of DNA extracted from calf thymus, *E. coli*, *H. influenzae* and *M. lysodeikticus* have been studied.

As the system possesses several constituents susceptible to change, various techniques have been utilised to determine the degree of evolution α of the transformation. For DNA the technique used was u.v. spectrophotometry with the property monitored being the optical density of the DNA. The concentration of the free Cu^{2+} ions can be readily determined as Cu^{2+} specific electrodes are available. Thus, for this species, the property monitored was the electrode potential.

B. VARIABLES

1. *DNA Source*

All other conditions kept constant, a mammalian DNA sample such as calf thymus renaturates at a rate the same order of magnitude as the renaturation rate of a bacterial DNA such as the DNA from *E. coli* (Fig. 1). Thus the complexity of the DNA does not seem to have any influence in this type of transformation. In contrast the GC composition is important. In fact the two types of DNA just mentioned, having similar GC composition renaturate at the same rate. DNA from *H. influenzae* and *M. lysodeikticus* having a lower and higher GC composition renaturate faster and slower respectively (Richard and Pacault, 1969; Richard 1970).

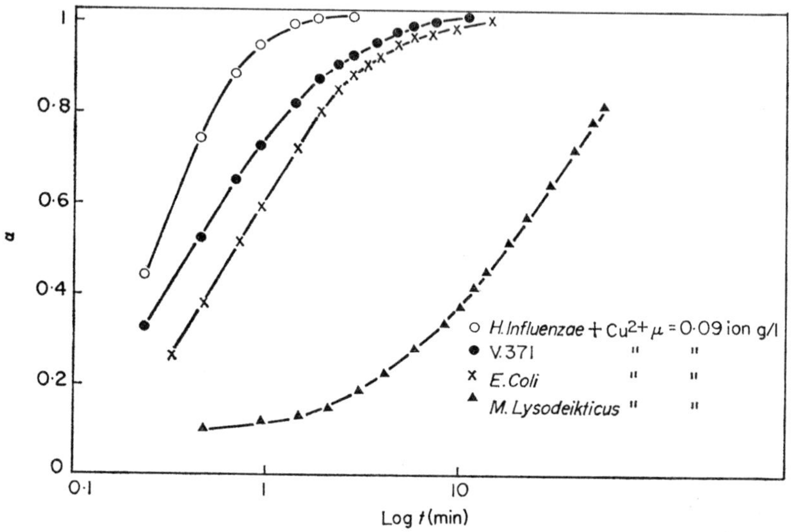

FIG. 1. *H. influenzae*, calf thymus (V371), *E. coli* and *M. lysodeikticus* DNA renaturation curves $\alpha = \phi(\log t)$.

2. *Denaturation Temperature T_1*

The renaturation rate decreases as the temperature of denaturation increases.

3. *Time θ of Waiting at the Temperature T_2*

After heating the DNA solution to the renaturation temperature T_1, it is tempered at the renaturation temperature T_2. There elapses a period of time θ between this tempering and the beginning of the renaturation triggered by the sharp increase of the ionic strength of the solution. The renaturation rate is a decreasing function of θ.

4. *Renaturation Temperature T_2*

In the range studied (16°C $< T_2 <$ 25°C), the variation of the rate constant k_T as a function of temperature more or less follows an Arrhenius rate law.† The activation energy is the order of 24 kcal (Richard and Pacault, 1968, 1969; Richard, 1970).

5. *DNA Concentration ("a" Remaining Constant)*

The renaturation rate decreases slightly with increasing concentration.

† Here the rate constant is the coefficient multiplied by the time in the evolution equation. The details of how this coefficient was determined will be described elsewhere (cf. IIC, 1).

6. *Ratio $a = Cu^{2+}/DNA-P$*

The renaturation rate diminishes as "a" increases. The rate constant k_a is such that log k_a is a linear function of $1/a$ (Richard and Pacault, 1968, 1969, 1970, 1971; Richard, 1970).

7. *Ionic Strength μ of Renaturation*

The greater the ionic strength of renaturation, the greater the renaturation rate. The coefficient k_μ is such that log k_μ is a linear function of $\sqrt{\mu}$.

It is quite possible that this enumeration of variables is not exhaustive but consists solely of those studied by Richard and Pacault (Richard and Pacault, 1968, 1969, 1970, 1971).

C. ANALYSIS OF THE KINETIC RESULTS

It was established (II) that the investigators studying DNA renaturation kinetics in the absence of ions such as Cu^{2+}, are endeavouring to deduce the transformation mechanism using formal kinetics. Richard and Pacault (1968, 1969, 1970, 1971) found that a kinetic law proposed by Pacault (1967, 1971) could be applied to a large number of complex systems and in particular DNA systems with Cu^{2+} ions present.

1. *Statement and Applications of the Kinetic Law*

Consider any given system in a state of equilibrium A, characterised by the values $p_1', p_2', p_3' \ldots p_i'$ of the independent intensive variables $p_1, p_2, p_3 \ldots p_i$ (temperature, pressure, mol fraction, etc.). By abruptly† changing the value of the variable p_1 from p_1' to p_1'' the investigator removes the system from its equilibrium state A and thus triggers the system's evolution towards a new equilibrium state B. This evolution is monitored by measuring a given property P (crystal parameters, index of refraction, optical density, electric or magnetic susceptibility, etc.) and the equation of evolution is of the form:

$$\alpha_P(t) = \phi_P\,(k_{p_1}\,t, p_1'', p_2', p_3' \ldots p_i')$$

where $p_2', p_3' \ldots p_i'$ remain constant throughout the change, t is the time measured by some periodic system (a chronometer for example), α_P is the extent of the evolution of the system measured by the relative change of the property P which characterises the system under fixed conditions determined by the variables $p_1, p_2, p_3 \ldots p_i$,

$$\alpha_P = \left(\frac{P_1 - P}{P_1 - P_2}\right)_{p_1, p_2, p_3 \ldots p_i)}$$

† Abruptly implies that the time required for the change from p_1' to p_1'' is negligible compared to the evolution time of the system.

The proposed law is the following:

The factor k_{p_1} multiplicative of the time parameter t is only dependent on the variable p_1, the abrupt variation of which triggers the evolution of the system. This is kept in mind on noting k_{p_1}.

As a consequence of the above, if several independent variables p_1, p_2, p_3 change value simultaneously and become p_1'', p_2'', p_3'', the evolution equation is of the form:

$$\alpha_P(t)=\phi(t.k_{p_1}.k_{p_2}.k_{p_3}, p_1'', p_2'', p_3'', p_i')$$

or

$$\alpha_P(t)=\phi\left(t\prod_j k_{p_j}, p_1'', p_2'', p_3'', p_i'\right)$$

With a view to checking the law it is convenient to plot $\alpha=\phi(\log t)$. The superposition of the curves obtained by a translation along the abscissa allows the determination of the magnitude, $\log k_{p_j}$, of the translation.

Examination of this law reveals that the variation of several parameters leaves the evolution expression unchanged. It emphasises an overall variation of the rate of evolution.

In fact, when each of the enumerated parameters is varied successively a large number of different curves $\alpha=\phi(t)$ are obtained. But applying the law in using a translation parallel to the abscissa restores all the curves to one unique curve governed by $\alpha=\phi(\log t)$.

It is thus evidently necessary to enumerate the evolution parameters of a system carefully in order to ensure their constancy during an experience. When this experimental control can be satisfactorily obtained, a single kinetic evolution curve of the system suffices for the determination of the general evolution expression.

2. *Illustration of the Law*

The kinetic law proposed by Pacault will be illustrated, by means of an example, e.g. the evolution curves of a sample of calf thymus DNA (V309) as a function of varying ionic strength μ, In Fig. 2 are shown the curves of the extent of evolution α as a function of time t: $\alpha=\phi(t)$. Figure 3 displays the function $\alpha=\phi(\log t)$. Figure 4 shows the result of the superposition of the preceding curves by translation parallel to the abscissa. Figure 5 shows the result of the superposition of the ensemble of the evolution curves of this sample as a function of ionic strength μ, renaturation temperature T_2, and the ratio $a=Cu^{2+}/DNA-P$. The equation shown is that which gives the rate constant k as a function of the variables μ, T_2 and a. Thus it can be seen how a large number of results can be synthesised giving one curve and one equation.

The system studied verifies the kinetic law (Pacault, 1967, 1971), whatever be the source of the DNA. Thus the 56 evolution curves of different DNA

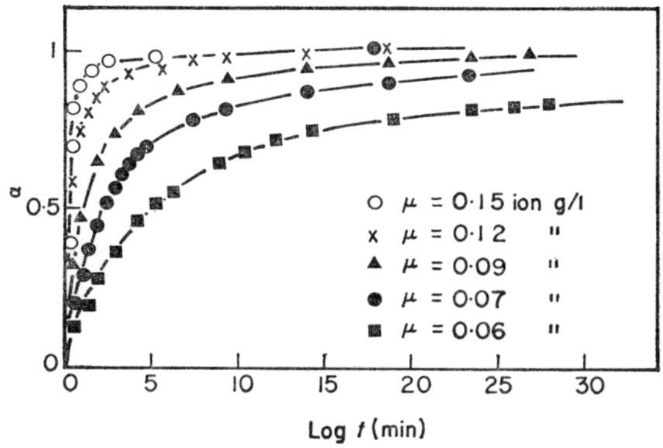

FIG. 2. Calf thymus DNA renaturation curves, $\alpha=\phi(t)$, for various values of the renaturation ionic strength.

types (calf thymus, *E. coli* and *H. influenzae*) as a function of various variables can be grouped into one curve (Fig. 6) with the deviations being within the limits of experimental errors.

3. *Evolution of a System Monitored by two Different Properties*

These two properties are the optical density of the DNA and the electrode potential of the Cu^{2+} ions. The system studied can be considered as consisting of two subsystems, one concerning the DNA and the other the Cu^{2+} ions. A comparison of the evolution of the two subsystems should give some indications of the nature of the transformation mechanism. It is established that the extent of evolution relative to the DNA α_d and the Cu^{2+} ions α_E is the

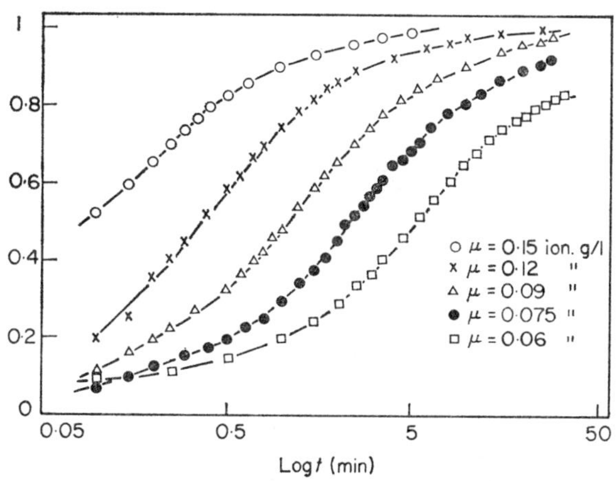

FIG. 3. Curves of Fig. 2 plotted against log t: $\alpha=\phi$ (log t).

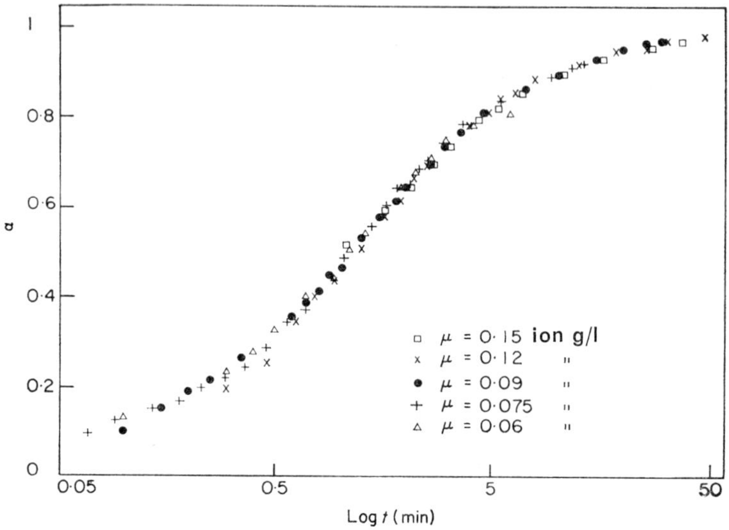

FIG. 4. Superimposition of the five curves of Fig. 3 obtained by a translation parallel to the abscissa.

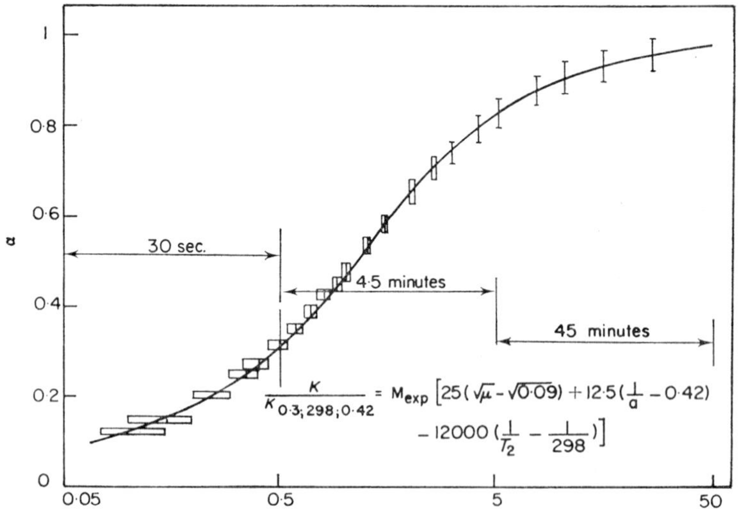

$$\frac{K}{K_{0\cdot3;\,298;\,0\cdot42}} = M_{\exp}\left[25(\sqrt{\mu}-\sqrt{0\cdot09}) + 12\cdot5\left(\frac{1}{a} - 0\cdot42\right) - 12000\left(\frac{1}{T_2} - \frac{1}{298}\right)\right]$$

FIG. 5. Result of the superimposition obtained by a translation parallel to the abscissa of fifteen curves, $\alpha = \phi(\log t)$ similar to those of Fig. 3. Equation giving the rate coefficient as a function of the variables ionic strength μ, ratio $a = Cu^{2+}/$ DNA-P and renaturation temperature T_2.

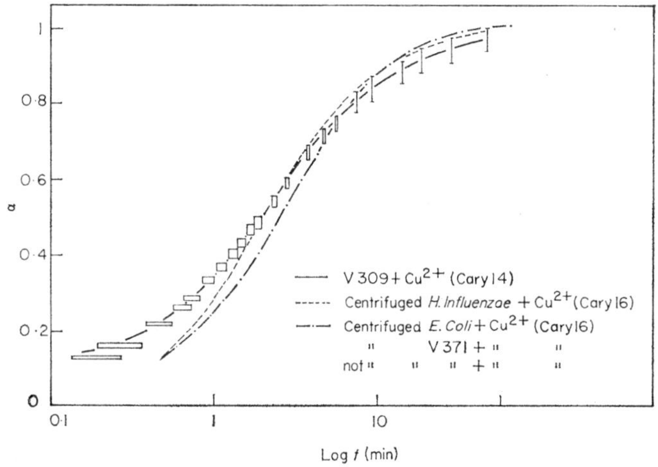

FIG. 6. Unique renaturation curve of various types of DNA (result of the superimposition of fifty-six curves).

same for the duration of the transformation (Richard, 1970; Richard and Pacault, 1971) (Fig. 7).

4. *Determination of the Kinetic Relation* $\alpha = \phi(t)$

A method of processing the kinetic curves is to determine the order of the reaction. In the presence of Cu^{2+} ions the DNA renaturation follows second-

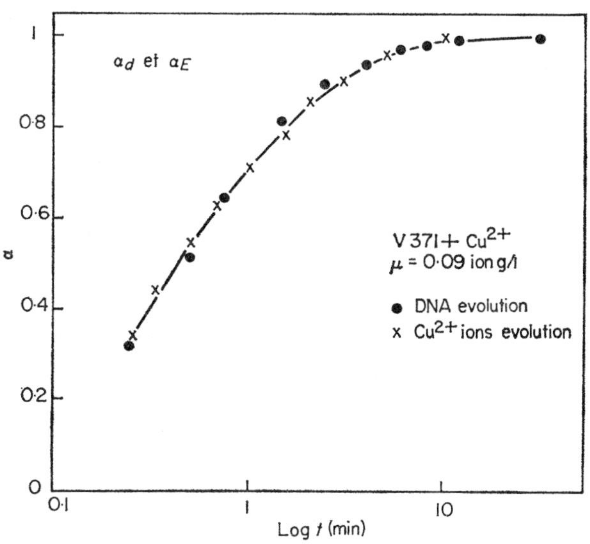

FIG. 7. Evolution curve $\alpha = \phi$ (log t) of the system determined from two different properties.

order kinetics when the extent of evolution is between 0·27 and 0·95. Keeping in mind these restrictions α can be written in the form:

$$\alpha = 1\bigg/\left(\frac{1}{k_2 t}+1\right)$$

But empirically, it was established that the experimental curves could also be described by the expression:

$$\alpha^2 = 1\bigg/\left\{1+\left(\frac{A}{t}\right)^k\right\}$$

Thus it would not be reasonable to draw firm conclusion concerning the mechanism on the basis of observed second-order kinetics.

D. RENATURATION MECHANISM OF DNA IN THE PRESENCE OF CUPRIC IONS

The phenomenological analysis consisting of classing the evolution curves of different systems according to their evolution expressions has shown that the systems studied evolve according to the same law (Fig. 6). This statement suggests the same mechanism of transformation. A comparison of the evolution rates of DNA from calf thymus, *E. coli*, *H. influenzae*, and *M. lysodeikticus* in the presence and absence of Cu^{2+} ions (Britten and Kohne) emphasises the importance of their role in the processes studied. It would seem that they (Cu^{2+} ions) behave as a screen between the complementary strands of the DNA, which would explain why the homogeneity of the DNA is a primary factor in the renaturation in their absence but not in their presence. As the GC composition has an appreciable influence it could be thought that the Cu^{2+} ions act at the GC base sites.

Finally the identity of the DNA evolution curves (Fig. 7) on the one hand and those of the cupric ions on the other indicates that the system is probably decomposable into two subsystems of which the rates of evolution are very different. One or the other method accounts for the slowest transformation, that is the evolution of the cupric ions system as without these ions the evolution of certain DNA types would not be possible.

Such are the indications resulting from the phenomenological analysis of the evolution curves of different DNA types in the presence of Cu^{2+} ions. Other methods of approach are possible. Their purpose is to determine the nature of the interactions between the DNA and the Cu^{2+} ions in order to explain the following facts:

The DNA is destabilised by the presence of an excess of Cu^{2+} ions in a solvent of weak ionic strength.

The DNA renaturation after thermal treatment at the temperature T_1 followed by cooling to the temperature T_2 is possible whatever be the source of the DNA.

The renaturation rate depends on the ionic strength of the triggering, on the ratio *a* and on the temperatures T_1 and T_2.

The model which best accounts for the ensemble of experimental results concerning the interactions between the Cu^{2+} ions and the DNA is due to Richard *et al.* (1973). A synthesis of the results obtained by mean of large number of physico-chemical methods (u.v. differential spectrophotometry, circular dichroism, optical rotary dispersion, melting curves, free Cu^{2+} ions dosage, determination of the number of disposable sites and of the constant of association of the ions on these sites) on studying DNA, poly dGdC, poly dAT, suggests the following model.

1. *Before Treatment at the Temperature T_1*

The DNA is in a solvent of weak ionic strength (Na_2SO_4) in the presence of an excess of Cu^{2+} ions ($CuSO_4$). In the neighbourhood of the phosphates of DNA there exists a strong concentration of Cu^{2+} ions, as under these conditions the competition between the Cu^{2+} ions and the Na^+ ions favours the Cu^{2+} ions. This strong concentration in Cu^{2+} around the macromolecule permits the occupation of all the disposable internal sites. These are of two sorts:

Interaction of a Cu^{2+} ion between two guanine plates belonging to two adjacent GC base pairs. This stabilizes the double helical structure but the number of sites is small ($\sim 5\%$).

PO^-–Cu–N_7 (guanine) bridging: this complex produces a local distortion and destabilises the corresponding GC base pairs.

The overall effect under these conditions is a destabilisation attributed to the occupation of the latter sites.

2. *During Treatment at Temperature T_1*

The adjacent GC base pairs are sufficiently stabilised by the Cu^{2+} ions intercalation complex so that the thermal treatment leaves them unchanged. These base pairs not having been destroyed constitute bonds permitting the complementary strands to remain face to face. This would explain why all the DNA can be renatured in the presence of Cu^{2+} ions.

In contrast the other GC base pairs destabilised by the PO^-–Cu^{2+}–N_7 bridging could separate and form zones of single strands. The same applies for the AT pairs. This single stranded zone formation frees the sites at G, C and A. There is a consequent redistribution of Cu^{2+} ions among the available sites. It is thus necessary to put forth the hypothesis that the excess of Cu^{2+} ions is sufficient so that the intercalation complexes are not destroyed at the expense of the new sites as the association constant is of the same order of magnitude.

3. *During Renaturation by Increasing the Ionic Strength*

When the ionic strength of the medium is increased, the previously established equilibrium is destroyed, i.e. that of:

$$DNA + Cu^{2+} \rightleftharpoons DNA - Cu^{2+}$$

In fact, an excess of Na^{+} ions relative to Cu^{2+} ions drives the Cu^{2+} ions from the neighbourhood of the phosphates of the macromolecule. In this manner, the complexes between the ions and the bases are destroyed by the parting Cu^{2+} ions from the phosphates to the solvent. The Cu^{2+} ions are progressively driven from the single strand zones and as a site becomes free the corresponding single strand zone undergoes instantaneous renaturation.

If it is accepted that the greater the quantity of Cu^{2+} ions and their affinity for the sites, the lesser the transformation rate, one can, on the basis of this model, explain the direction of the variation of the renaturation rate as a function of each one of the variables studied:

GC composition of DNA: the affinity between Cu^{2+} ions and guanine is greater than that between Cu^{2+} ions and adenine. A DNA rich in GC complexes more extensively with Cu^{2+} ions and the renaturation rate is consequently slower.

Denaturation temperature T_1: the higher the temperature, the more the single strand structure is favoured as the number of accessible sites increases with temperature. This has the result of diminishing the renaturation rate.

Renaturation temperature T_2: if an Arrhenius type behaviour is followed, the rate must increase with T_2.

Ionic strength of renaturation: the greater the ionic strength, the more the decomplexation reaction of the Cu^{2+} ions is favoured. An increase in the rate of decomplexation translates as an increase in the renaturation rate.

Ratio $a = Cu^{2+}/DNA - P$: increasing the ratio favours the complexation of Cu^{2+} ions which results in a diminution of the renaturation rate.

The directional variation of the transformation rate anticipated by the model is in effect that observed experimentally.

IV. Conclusion

The kinetics of DNA renaturation in the absence of Cu^{2+} ions has been studied under conditions consistent with those in living systems. Only the denaturing treatment using temperature elevation is incompatible. Analysis of the results on the basis of the reaction order has given rise to a large number of inconsistent results requiring various assumptions to explain them. Biologists have tried, in such a manner, to understand the synthesis of DNA.

Introduction of a large quantity of Cu^{2+} ions considerably removes the conditions from those observed in living systems. Although a second order expression is observed, as with the preceding case, a certain number of behavioural differences seem to eliminate the possibility of the same mechanism being operative in the presence and absence of Cu^{2+} ions. The phenomenological analysis permits the consideration of Cu^{2+} ions and the DNA as an entity DNA–Cu^{2+}, behaving differently than DNA alone. The model proposed by Richard, Daune and Schreiber provides an explanation for this.

References

Blake, R. D. and Fresco, J. R. (1966). *J. molec. Biol.* **19**, 145.

Britten, R. J. and Kohne, D. E. (1968). *Science N.Y.* **161**, 529.

Cavalieri, L. F., Small, Th. and Sarkar, N. (1962). *Biophys. J.* **2**, 339.

Flory, P. J. (1961). *J. Polymer Sci.* **49**, 105.

Inman, R. B., Baldwin, R. L. (1964). *J. molec. Biol.* **8**, 452.

Kallenbach, N. R., Crothers, D. M. and Mortimer, R. G. (1963). *Biochem. biophys. Res. Commun.* **11**, 213.

Marmur, J. and Doty, P. (1961). *J. molec. Biol.* **3**, 585.

Marmur, J., Rownd, R. and Schildkraut, C. L. (1963). "Progress in Nucleic Acid Research". Academic Press, New York and London.

Pacault, A. (1967). Nobel Symposium. Stockholm.

Pacault, A. (1971). Conférence. Académie des Sciences, Dusseldorf.

Richard, H. and Pacault, A. (1968). *Bull. Soc. Chim. biol.* **50**, 417.

Richard, H. and Pacault, A. (1969) *C.r. hebd. Séanc. Acad. Sci. Paris* **268**, 1650.

Richard, H. (1970). Thèse Bordeaux.

Richard, H. and Pacault, A. (1971). *Biopolymer* **10**, 2299.

Richard, H., Daune, M. and Schreiber, J. P. (1973) *Biopolymer* **12**, 1.

Ross, P. D. and Sturtevant, J. M. (1960). *Proc. natn. Acad. Sci. U.S.A.* **46**, 1360.

Ross, P. D. and Sturtevant, J. M. (1962). *J. Amer. chem. Soc.* **84**, 4503.

Saunders, M. and Ross, P. D. (1960). *Biochem. biophys. Res. Commun.* **3**, 314.

Studier, F. W. (1969). *J. molec. Biol.* **41**, 199.

Subirana, J. A. and Doty, P. (1966). *Biopol.* **4**, 171.

Thrower, K. J. and Peacoke, A. R. (1966). *Biochim. biophys. Acta* **119**, 652.

Wetmur, J. G. and Davidson, N. (1968). *J. molec. Biol.* **31**, 349.

Wetmur, J. G. (1971). *Biopol.* **10**, 601.

CHAPTER 18

Interactions in Nucleic Acids

M. LENG, M. DOURLENT AND C. HÉLÈNE

Centre de Biophysique Moléculaire, Orleans, France

Knowledge of the structure and conformation of nucleic acids is an essential step in our understanding of their biological activity. The double helical character of DNA and its tertiary structure in nucleoprotein complexes control duplication and transcription processes. A particular secondary structure of some regions of DNA molecules has been assumed to explain the selective binding of repressor molecules. The tertiary structure of transfer RNAs is essential for their selective recognition by aminoacyl *t*RNA synthetases. In ribosomes, the structure of RNA is an important factor for the association with proteins.

Elucidation of the structure of nucleic acids rests mainly upon two lines of experiments:

i. Studies of fibres by X-ray diffraction techniques has allowed a relatively precise definition of the conformation of certain types of nucleic acids; up to now the structure of *t*RNA has not been solved by X-ray crystallography.

ii. Studies in solution by various physical techniques have led to characterisation of the secondary and tertiary structures of different kinds of nucleic acids. Much has been learnt from studies of polynucleotides in coil, single-stranded or double-stranded forms.

The structure of nucleic acids and polynucleotides is contributed to by the restriction imposed by the permitted conformation of the phosphate-ribose backbone and by direct interactions between purine and pyrimidine bases. These interactions can be investigated by the study of monomers (bases, nucleosides, nucleotides) in aqueous and non-aqueous solvents and also in crystals of purine and pyrimidine derivatives. This will represent the first point of our discussion. The role played by the phosphate-ribose backbone can be conveniently studied in polynucleotides either single-, double- or triple-stranded. This will be discussed in the particular cases of poly A, poly U and their complexes poly A·poly U and poly A·2 poly U.

Although the intrinsic conformation of nucleic acids under well-defined experimental conditions is important in the understanding of their properties, the changes in conformation induced by the binding of other molecules appear essential for the behaviour of the complexes. We will not attempt to discuss the conformational changes induced by the binding of macromolecules such as enzymes or proteins but rather focus our attention on the binding of small organic molecules and in particular of dyes which can be considered as models for the binding properties of other molecules such as aromatic hydrocarbons or aromatic amino acids.

I. Monomer Interactions

A. INTERACTIONS BETWEEN BASES AND BASE-PAIRS IN CRYSTALS

1. *Base Crystals*

Crystals from bases, nucleosides or nucleotides can be grown from aqueous solutions or from aqueous mixtures containing ethanol or dimethylsulphoxide. The data obtained from X-ray crystallographic studies have been reviewed (Davies, 1967; Haschemeyer and Rich, 1967; Sasisekharan *et al.*, 1967; Ts'o, 1970). Both hydrogen bonding and stacking interactions are observed in these crystals. The bases are mostly planar with displacements usually much less than 0·1 Å.

In crystals of nucleosides or nucleotides, it is generally observed that four atoms of the furanose ring are coplanar. The fifth one (C(2′) or C(3′)) is displaced from this plane by 0·5 to 0·6 Å. The out of plane atom may be displaced on the same side as the C(5′) atom (*endo* conformation) or on the opposite side as compared to the C(5′) atom (*exo* conformation). The furanose conformation is very sensitive to the environment and may vary in different crystals of the same compound.

Crystals may be hydrated or anhydrous (even when grown from aqueous solution). Both anhydrous and hydrated crystals of the same compound may be obtained for example with cytosine (Jeffrey and Kinoshita, 1963; Barker and Marsh, 1964). Bond distances and bond angles are very similar in both cases. In the hydrated crystals, the hydrogen bonded structure is perturbed by water molecules through competition for the hydrogen-bond donor and acceptor groups of the bases. An interesting feature found in crystals of adenosine-3′-phosphate dihydrate is the presence of C(8)—H. . .O (water) hydrogen bond with H. . .O and C. . .O distances of 2·24 Å and 3·26 Å respectively (Sundaralingam, 1966). Evidence for such a C—H. . .O bonding was also obtained from nuclear magnetic resonance (n.m.r.) studies of purine in several solvents (Hruska *et al.*, 1968). An hydrogen bond formation involving the C(6)H of pyrimidine derivatives has been suggested (Ts'o, 1970).

Different crystallographic forms of the same compound may be obtained under certain experimental conditions. For example cytidine-3′-monophosphate (CMP) crystallises into an orthorhombic form and a monoclinic form (Bugg and Marsh, 1967). The molecular structure and conformation of CMP is practically the same in these two anhydrous forms, the main difference arising from changes in hydrogen bonding and packing of the molecules.

Relationships between conformation and bond angle can be found in the review by Sasisekharan *et al.*, (1967).

As far as rotation around the C(1′)–N(9) or N(1) bond is concerned, the results of X-ray studies on 16 pyrimidine derivatives and 11 purine derivatives have been tabulated by Ts'o (1970). The pyrimidine compounds are all observed to be in the *anti* conformation whereas the energy barrier between the *syn* and *anti* conformations appears to be much smaller for purine derivatives, some of them preferring the *syn* conformation (guanosine for example) (Haschemeyer and Rich, 1967; Sundaralgrigam and Jensen, 1965).

2. *Interactions between Base Pairs in Crystals*

Since the first successful co-crystallisation of 9-methyladenine and 1-methylthymine by Hoogsteen (1959, 1963), several complexes of bases have been investigated by X-ray crystallography.

These intermolecular complexes are essentially planar and are formed only with the complementary bases of nucleic acids.

In all the complexes between adenine and uracil derivatives hydrogen-bonding involves N(7) and not N(1) of adenine (Hoogsteen, 1959, 1963; Mathews and Rich, 1964; Katz *et al.*, 1966; Baklagine *et al.*, 1966; Tomita *et al.*, 1967; Haschemeyer and Sobell, 1965). Two types of structures have been found: the "Hoogsteen" and the "reversed Hoogsteen" structures, shown in Fig. 1. The relationship between these two base-pairing schemes is a rotation of the pyrimidine by 180° around the N(3)–C(6) line which substitutes the C(4)=O ("Hoogsteen" type) with the C(2)=O ("reversed

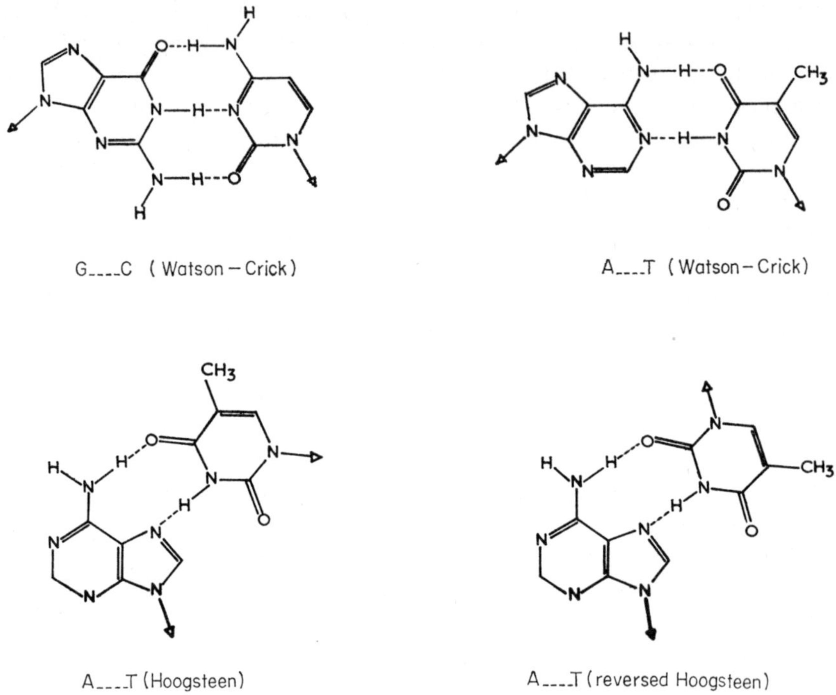

FIG. 1. Different types of base pairing schemes.

Hoogsteen" type). None of these A–U or A–T crystalline complexes corresponds to the Watson–Crick base-pairing scheme.

On the opposite, all the G–C complexes investigated have the Watson–Crick type of hydrogen-bonding scheme (Fig. 1) with 3 hydrogen bonds (Sobell *et al.*, 1963; O'Brien, 1963, 1966; Haschemeyer and Sobell, 1965).

From all these studies it can be concluded that although hydrogen-bonding is essential in determining the selectivity of interactions, the vertical packing of the complexes could be an important parameter in choosing between different types of hydrogen-bonded complexes. Base stacking in crystals of nucleic acid derivatives has been reviewed recently (Bugg *et al.*, 1971). Only partial base overlap is observed in most of the known structures, polar substituents being usually positioned over the ring of an adjacent base. These interactions are different from the ring interactions in non-polar aromatic compounds but similar to those found in the crystal structures of other polar aromatic compounds. In a recent paper, Arnott *et al.*, (1969) have pointed out that although there is certainly overlap of the base-pairs as a whole in DNA and RNA, there is relatively little overlap of the heterocyclic rings themselves. Theoretical calculations have compared the relative importance of electrostatic forces (calculated in the dipole–dipole or monopole–monopole approximation), polarisation and dispersion forces in the stacking interaction (De Voe and Tinoco, 1962; Claverie *et al.*, 1966). From the comparative studies of Bugg *et al.* (1971), it appears that dipole-induced dipole (polarisation) forces are largely responsible for base stacking in crystals.

B. HYDROGEN-BONDING BETWEEN NUCLEIC ACID DERIVATIVES IN NON-AQUEOUS SOLVENTS

The formation of hydrogen bonds between complementary bases of nucleic acids is not restricted to crystals. In solvents which have no or weaker hydrogen-bonding capacity than water, competition between solute and solvent for interaction should be decreased. Using purine and pyrimidine derivatives soluble in chloroform or carbon tetrachloride, several groups of investigators have demonstrated the formation of hydrogen-bonded dimers by i.r. spectroscopy (Hamlin *et al.*, 1965; Kyogoku *et al.*, 1966, 1967a, b; Kuchler and Derkosch, 1966; Pitha *et al.*, 1966; Miller and Sobell, 1967).

Infrared measurements were carried out mainly in the region of NH and NH_2 stretching vibrations (2800–3500 cm^{-1}). In some cases, the carbonyl stretching vibration (1700 cm^{-1}) has also been used. Self-association through hydrogen-bonding of each of the compounds investigated is clearly demonstrated by the dependence of the i.r. spectrum upon concentration. When the spectra of mixtures of adenine and uracil derivatives, or of guanosine and cytidine derivatives are compared to the separated components at the same concentration, a series of pronounced bands due to hydrogen-bonding appears. These bands can be used to determine the stoichiometry of the complexes by the method of continuous variation (Job, 1928). In most cases, 1 : 1 complexes are formed.

The i.r. data do not give direct information concerning the configuration of the dimers in solution. They are compatible with the configurations obtained from X-ray diffraction studies of co-crystals (see above).

Only the interaction between complementary bases of nucleic acids can be detected by i.r. spectroscopy in organic solvents. The same conclusion is drawn from p.m.r. studies (Katz and Penman, 1966; Shoup *et al.*, 1966; Newmark and Cantor, 1968). The n.m.r. spectrum of the base protons provides a sensitive tool for investigating their local magnetic environment. If hydrogen bonds are formed, the large decrease in magnetic shielding of the hydrogen atoms involved in bonding leads to a downfield shift of their resonances. Due to fast chemical exchange between bondforming hydrogen atoms and water protons, it is not possible to use H_2O (or D_2O) as a solvent. H_2O can only be used if these hydrogen atoms exchange slowly enough with the solvent as observed for amino groups (Razka and Kaplan, 1972) in double-stranded regions of *t*RNAs (Kearns *et al.*, 1971).

P.m.r. studies of base-pair formation were carried out in $CDCl_3$, in dimethylsulphoxide (DMSO) and its mixtures with dimethylformamide (DFM) or benzene. DMSO is a strong H-bond acceptor and extensive bonding of NH and NH_2 groups to this solvent has been indicated by the temperature dependence of NH and NH_2 resonances and the comparison of chemicals shifts in DMSO and $CDCl_3$ (Katz and Penman, 1966). However, large downfield shifts of the H-bonding protons are observed when cytidine (and only cytidine) is added to guanosine. Cytidine manifests no interactions with any of the nucleosides except guanosine. The p.m.r. results support a Watson–Crick GC pairing since both amino groups and the NH of guanosine are involved (Shoup *et al.*, 1966).

No interaction is observed between adenosine and uridine in DMSO probably because the competitive interaction with the solvent is too strong in this case. However an interaction between uracil and adenine derivatives can be demonstrated by p.m.r. studies in $CDCl_3$ or in mixtures of DMSO and benzene. In every case only one resonance peak is observed for the free and bonded protons indicating that fast exchange between the free and complexed form does occur. Kinetic studies of hydrogen-bonding between 1-cyclohexyluracil and 9-ethyl adenine in $CHCl_3$ using ultrasonic absorption measurements have shown that complex formation is controlled by diffusion ($k_a = 4 \times 10^9$ M^{-1} $\sec^{-1}$ at 20°C) with a dissociation rate constant of $3{\cdot}2 \times 10^7$ $\sec^{-1}$ (Hammes and Park, 1968).

Although p.m.r. spectra are much simpler to analyse than i.r. spectra, i.r. measurements require much less concentrated solutions. However one of the most important differences between i.r. and p.m.r. is that every species present in the solution has its own i.r. absorption irrespective of its lifetime which is not the case for hydrogen-bonded complexes in p.m.r. spectroscopy.

Infrared p.m.r. data were used to calculate the association constants for the formation of hydrogen-bonded complexes and to obtain thermodynamic parameters (enthalpy and entropy change). In Table I are summarised the results for 9-ethyladenine and 1-cyclohexyluracil in $CDCl_3$. Complete results for different uracil and adenine derivatives have been published (Kyogoku *et al.*, 1967b).

The data concerning G–C base-pair formation are less abundant. Association constants in $CHCl_3$ range from 10^4 (Kyogoku *et al.*, 1966) to 10^5 M^{-1} (Pitha *et al.*, 1966). In DMSO, a value around 4 M^{-1} has been found by p.m.r. studies (Newmark and Cantor, 1968) (Table I).

Theoretical calculations of the different forces contributing to the stability of hydrogen-bonded base pairs have been carried out (Nash and Bradley, 1966; Bradley and Nash, 1968; Pullman *et al.*, 1966a, b; Caillet, 1968)

TABLE I

Association constants and thermodynamic parameters for formation of hydrogen-bonded complexes in $CDCl_3$

	A–A*	U–U*	A–U*	G–C†
K (M^{-1}) at 25°C	3·1	6·1	100	3·6–4·0
ΔH (kcal/mol)	4·0	4·3	6·2	5·8
ΔS (eu)	11·4	11·0	11·8	16

* Kyogoku *et al.*, 1967 (in $CDCl_3$).
† Newmark and Cantor, 1968 (in DMSO).

and have been successful in explaining the observed specificity of interactions between A and U and between G and C. There has been also calculations on the relative stability of different configurations of the protons in the hydrogen bonds (Rein and Harris, 1966; Lunell and Sperber, 1967).

Base-pair formation in $CHCl_3$ is accompanied by a change (mostly hypochromism) in the absorption spectrum (Thomas and Kyogoku, 1967; Gratzer and McClare, 1967). The magnitude of the change is between 5% and 20% depending on the base pair and the wavelength at which the change is measured. This result shows that hypochromicity can be induced not only by stacking interactions (as assumed in double-stranded nucleic acids) but also by the formation of coplanar hydrogen-bonded pairs.

C. INTERACTIONS BETWEEN NUCLEIC ACID DERIVATIVES IN AQUEOUS SOLUTIONS

Self-association of purine and pyrimidine derivatives in aqueous solutions has been demonstrated by different methods: vapour pressure osmometry (Ts'o *et al.*, 1963; Ts'o and Chan, 1964; Broom *et al.*, 1967; Ts'o *et al.*, 1969; Solie and Schellman, 1968; Magar and Steiner, 1970; Helmkamp and Kondo, 1968), changes in the apparent molar volume upon dilution (Kasarda, 1970), sedimentation equilibrium (Van Holde and Rossetti, 1967; Solie and Schellman, 1968), calorimetric techniques (Gill *et al.*, 1967; Farquhar *et al.*, 1968), absorbance and ORD† (Gellert *et al.*, 1962; Solie and Schellman, 1968; Sarkar and Yang, 1965) and p.m.r. (Chan *et al.*, 1964; Schweizer *et al.*, 1968; Broom *et al.*, 1967; Schweizer *et al.*, 1968; Helmkamp and Kondo, 1968; Danner and Helmkamp, 1971; Jardetzky, 1965).

Vapour pressure osmometry data clearly indicate that aqueous solutions of bases and nucleosides are far from ideal. Osmotic and activity coefficients are much below unity, providing evidence that purine and pyrimidine derivatives interact extensively in aqueous solutions. Different models have been used to account for the experimental data obtained by vapour pressure

† Optical Rotatory Dispersion.

osmometry and sedimentation equilibrium. Formation of dimers only is not compatible with thermodynamic data (Ts'o *et al.*, 1963). Thus association must involve a higher degree of polymerisation. A model assuming that the association constant is the same for the different steps of the association process fits the osmometry and sedimentation data (Ts'o *et al.*, 1963; Solie and Schellman, 1968). This association constant is obtained from eqn (1) where m is the molality of the solute and ϕ the osmotic coefficient

$$K = 1 - \phi/m\,\phi^2 \tag{1}$$

The degree of association was calculated to be higher than 5 for 6-methylpurine (Ts'o and Chan, 1964). In the case of 5-bromo uridine it was found necessary to introduce two equilibrium constants, $K_1 = 1{\cdot}0\ \text{M}^{-1}$ for the first step and $\bar{K} = 2{\cdot}9\ \text{M}^{-1}$ for the successive steps (up to the tetramer). Values of association constants for several compounds, calculated assuming identical equilibrium constants for every association step, are given in Table II. Enthalpy and entropy values for self-association have been calculated by Gill *et al.* (1963) and Farquhar *et al.* (1968) from microcalorimetry and osmometry data. More recently, Magar and Steiner (1970) have proposed to analyse the results of pressure vapour osmometry measurements without the restrictive assumption of identical constants for successive steps. Least square analysis and linear programming fit show that self-association of inosine, cytidine and uridine could be limited to trimers in the concentration

TABLE II

Thermodynamic parameters for self-association of purine and pyrimidine derivatives in aqueous solution (data from Gill et al., 1963; *Ts'o* et al., 1963; *Farquhar* et al., 1968; *Broom* et al., 1967)

Compound	K (M^{-1}) (25°C)	ΔH (kcal/mol)	ΔS(eu)
Purine	2·1	−4·2	−13
6-Methylpurine	6·7	−6·0	−16
Caffeine	13	−3·4	−6
Cytidine	0·87 (1·26)† (7·9)	−2·8 (−3·2)*	−10
Uridine	0·16 (1·18)† (2·6)	−2·7 (−2·4)*	−10
Thymidine	0·91	−2·4	−9
Adenosine	4·5	—	—
N-6-Dimethyladenosine	22·2	—	—

† Values of the two successive association constants and enthalpy of first step calculated by Magar and Steiner (1970).

range investigated (0·1 to 0·75 M for cytidine and uridine, 0·05 to 0·2 M for inosine) (see Table II).

The mode of self-association of purine derivatives has been extensively studied by proton magnetic resonance. Due to the large magnetic anisotropy of aromatic rings, stacking interactions are expected to lead to upfield shifts of the ring proton resonances while hydrogen-bonding (in-plane) interactions should lead to downfield shifts. The p.m.r. spectra of purine have been studied by Chan *et al.* (1964) over the concentration range 0·05–1 M. Proton resonances in purine are all shifted to higher fields when the concentration increases. Increasing the temperature (which is expected to decrease the interactions) leads to downfield shifts (Helmkamp and Kondo, 1967). Protonation of purine eliminates the concentration effect (Chan *et al.*, 1964). Extensive studies of purine nucleosides (Broom *et al.*, 1967) demonstrated that an increase of the concentration leads to upfield shifts of the base protons (particularly H2) and of the anomeric proton H–1′, whereas the resonances of the other ribose protons are less concentration dependant (especially H–5′). This indicates that adenine nucleoside interactions are preferentially localised at the purine base and involves vertical stacking of the aromatic rings. Several preferred average orientations of the bases have been proposed to account for the p.m.r. results (Broom *et al.*, 1967; Helmkamp and Kondo, 1967). Successful correlation between p.m.r. and osmotic data has been made for purine self-association under the assumption that the chemical shifts of associates whose degree of polymerisation is higher than 3 are identical.

Recently, p.m.r. data for purine self-association were analysed without relying upon the osmotic data (Dimicoli and Helene, 1973). In this model, identical equilibrium constants are assumed for the successive association steps and the magnetic anisotropy of nearest neighbours only is taken into account. Under these conditions the chemical shift of an n-mer (B_n) is given by $\sigma_{B_n}=\{2(n-1)\,\sigma_{B_2}-(n-2)\,\sigma_B\}/n$. This leads to eqn (2) where $\delta\sigma=\sigma-\sigma_B$ and $\delta\sigma_{B_2}=\sigma_{B_2}-\sigma_B$, σ is the chemical shift of a purine proton at concentration B_0, σ_{B_2} is the average chemical shift in a dimer and σ_B is the chemical shift of the same proton extrapolated to 0 concentration.

$$\left(\frac{\delta\sigma}{B_0}\right)^{1/2}=\left(\frac{K}{2\,\delta\sigma_{B_2}}\right)^{1/2}(2\delta\sigma_{B_2}-\delta\sigma) \qquad (2)$$

Plots of $(\delta\sigma/B_0)^{1/2}$ versus $\delta\sigma$ are straight lines for the three protons of purine (Fig. 2). The calculated value of K is $2{\cdot}2 \pm 0{\cdot}2$ M^{-1} in good agreement with that determined from osmotic data (2·1 M^{-1}, Table II).

The self-association of some purine derivatives in aqueous solutions (3′–GMP, 5′–GMP, iso-guanosine) leads to gel formation which involves both hydrogen-bonding and stacking interactions (Gellert *et al.*, 1962; Sarkar and Yang, 1965; Ravindranathan and Miles, 1965). This represents

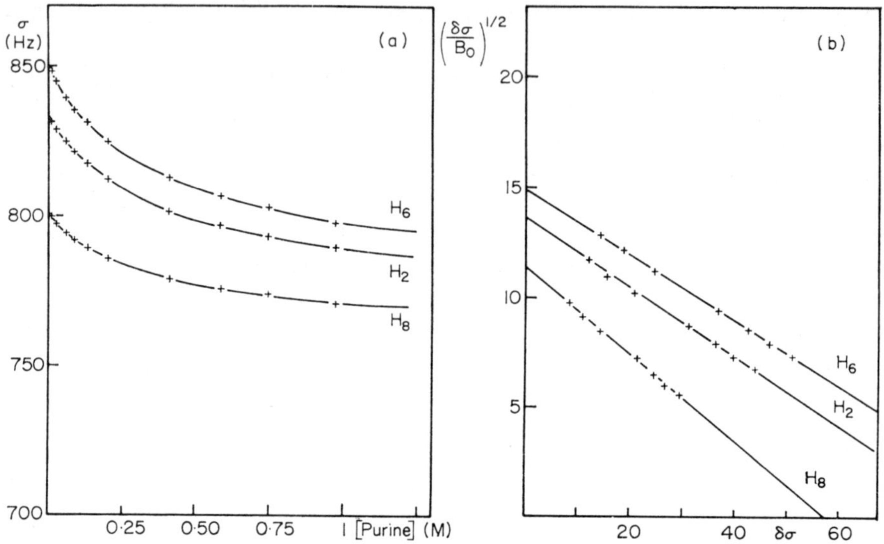

FIG. 2. (a) concentration dependence of the chemical shifts of different protons of purine. (b) analysis of the p.m.r. data for purine according to eqn (2) (see text and Dimicoli and Hélène, 1972).

the only cases of self-association in aqueous solutions where hydrogen-bonding has been shown to be involved. For most nucleosides or nucleotides investigated, a stacking of the bases is the only interaction which has been demonstrated. Methylation of the hydrogen-bond donor sites (such as in N-dimethyl adenosine) enhances the association constants even though hydrogen-bonding is prevented.

The association of pyrimidine nucleosides in aqueous solutions cannot be studied by p.m.r. The ring current magnetic anisotropy of the pyrimidine ring is too small to induce upfield shifts of the proton resonances of neighbouring molecules although these compounds have been shown to associate by osmotic measurements.

Heterologous associations between purine and pyrimidine derivatives have been investigated by vapour pressure osmometry (Solie and Schellman, 1968; Steiner, 1971; Magar *et al.*, 1971), solvent partition (Gratzer, 1969), the phase-solubility method (Ts'o *et al.*, 1963; Nakano and Igarashi, 1970; Gill *et al.*, 1963), p.m.r. (Danner and Helmkamp, 1971; Schweizer *et al.*, 1965; Dimicoli and Hélène, 1973). The most important result is that the tendency for interaction of bases and nucleosides in aqueous solutions decreases in the series: purine–purine > purine–pyrimidine > pyrimidine–pyrimidine.

One of the difficulty encountered in the quantitative study of heterologous associations comes from self-association of both partners. Equilibrium constants have often been determined assuming 1 : 1 complex formation (Nakano and Igarashi, 1970) or assuming that each aggregate of one compound (B) combines with free monomeric (A) as if it were not self-associated (Ts'o *et al.*, 1963). More recently, Steiner (1970) presented a general method for obtaining consecutive association constants from colligative data (such as vapour pressure osmometry) obtained when the two types of interacting molecules undergo self-association. This method was used to determine the association constants for purine–uridine (Steiner, 1971) and inosine–cytidine (Magar *et al.*, 1971) interactions which the conclusion that complex formation extends beyond the 1 : 1 species. Theoretical models have also been proposed to analyse p.m.r. data on heterologous associations and applied to purine–cytosine and purine–caffeine (as well as purine–5-hydroxyindole) interactions (Dimicoli and Hélène, 1973).

Association constants for heterologous associations are of the same order of magnitude as those determined for self-association ($\leqslant 10\ \text{M}^{-1}$, except for alkylxanthines which give higher association constants).

Interaction between pyrimidine nucleosides and purine nucleosides is demonstrated by important upfield shifts of the pyrimidine base and H–1′ protons. The ribose proton resonances are less and less shifted as one proceeds from H–1′ to H–5′, as already observed when purine nucleosides associate. This indicates that association is due mainly to base–base stacking interaction. Evidence for insertion of pyrimidine bases into purine aggregates is provided by the downfield shifts of the purine proton resonances in the presence of pyrimidine nucleosides, the magnetic anisotropy of the pyrimidine ring being far less important than that of purine. No evidence for hydrogen-bond formation between complementary bases of nucleic acids has been found by Raman spectroscopy in aqueous solutions (Lord and Thomas, 1967).

Theoretical calculations of the different contributions to the stacking interactions between the bases indicate that the London dispersion energy is more important than polarisation and permanent dipole–dipole (monopole–monopole) energies. In agreement with this finding, there does not seem to be any correlation between the dipole moment of the bases and their tendency to associate (Ts'o, 1970). The role of the solvent in complex formation between nucleosides in aqueous solvents has not been clearly elucidated yet. Self-association of bases and nucleosides takes place only in water and not in organic solvents. A hidden positive entropy due to the solvent has been suggested to contribute to stacking interactions in water (see Ts'o, 1970).

Stacking interactions between bases in dinucleotides have been extensively studied by ORD, circular dichroism (CD) and p.m.r. (see Warshaw and

Tinoco, 1966; Brahms *et al.*, 1967; Chan and Nelson, 1969; Bangerter and Chan, 1969; Ts'o, 1969).

Stacking interactions involving purine or pyrimidine derivatives and aromatic amino acids have been recently demonstrated. Tyrptophan forms complexes with purine and pyrimidine derivatives either in frozen dilute aqueous solutions (Montenay-Garestier and Hélène, 1971) or in more concentrated fluid aqueous solutions (Dimicoli and Hélène, 1971). Proton magnetic resonance data indicate that the purine or pyrimidine ring and the indole ring are stacked (Dimicoli and Hélène, 1971, 1973; Hélène *et al.*, 1971a). The aromatic ring of tyrosine has also been shown to form stacked complexes with the purine or pyrimidine rings (Hélène *et al.*, 1971b).

D. ASSOCIATION OF NUCLEOSIDES OR NUCLEOTIDES INDUCED BY INTERACTIONS WITH POLYMERS

The tendency of purine nucleosides or nucleotides to associate in aqueous solutions is greatly enhanced in the presence of the complementary polynucleotide.

The interaction of guanosine-5′-mononucleotides with poly C in aqueous solutions was first investigated by Howard *et al.* (1964) using i.r. spectroscopy. The formation of double-stranded and triple-stranded structures with cooperative melting was reported. Ts'o and Huang (1968) later reported that complex formation between GMP, d-GMP or GTP and poly C as well as between AMP, dAMP, ATP or dATP and poly U is accompanied by a phase transition. Insoluble complexes are formed with a definite stoichiometry 2U to 1A, 1C to 1G at low concentrations or 2C to 1G at high concentrations. No soluble complex could be detected by ORD, sedimentation or p.m.r. (Ts'o and Schweizer, 1968) though insoluble complexes are formed at lower temperatures (precipitation or gel formation).

The specific and cooperative interaction of nucleosides with polynucleotides was investigated by various techniques: equilibrium dialysis, optical rotatory dispersion, ultracentrifugation, enzymatic degradation (Huang and Ts'o, 1966; Howard *et al.*, 1966) proton magnetic resonance (T'so and Schweizer, 1968; Bangerter and Chan, 1968) and calorimetric measurements (Scruggs and Ross, 1970). The formation of adenosine–poly U complexes with a stoichiometry of 2U/1A is characterised by a strongly cooperative binding isotherm as determined by equilibrium dialysis (Huang and Ts'o, 1966). No binding of cytidine, inosine or N(6)-methyladenosine could be detected. Increase in both the sedimentation coefficient and the viscosity of poly U are observed upon binding of adenosine. The thermal denaturation of the complexes followed by ORD or i.r. spectroscopy is characterised by a sigmoidal curve, indicative of cooperative melting of an ordered structure.

P.m.r. investigations demonstrate that adenine–uracil hydrogen-bonding and adenine–adenine stacking take place in the three-stranded complex of poly U and adenosine. Above the melting temperature adenosine was shown to intercalate in a non-cooperative way between uracil residues (Bangerter and Chan, 1968).

Interaction between poly C and guanosine was demonstrated by Howard *et al.* (1966) using i.r. spectroscopy.

Binding of caffeine and guanosine to the poly U–adenosine complexes takes places by co-stacking with adenosine since no binding is observed in the absence of adenosine (Pitha *et al.*, 1968; Schmitz and Schurr, 1971). Such co-stacking is not observed with cytidine as expected since purine–pyrimidine interactions are weaker than purine–purine interactions.

The main conclusions which can be drawn from the investigation of several systems are:

i. Purine monomers form complexes with pyrimidine polynucleotides but no pyrimidine monomer–purine polynucleotide complex can be detected.

ii. The binding is specific, taking place only with the complementary bases pairs of nucleic acids and thus involving hydrogen-bonding interactions.

iii. The binding is cooperative as a result of stacking interactions between purine monomers (which are stronger than those between pyrimidine monomers as shown in the preceding section).

The association of nucleotides in solutions can be increased in the presence of polynucleotides but also in the presence of basic polypeptides such as poly-L-arginine or poly-L-lysine.

The binding of nucleotides by poly-L-arginine and poly-L-lysine was investigated by equilibrium dialysis (Wagner and Arav, 1968; Arav and Wagner, 1970). In several cases, especially with adenosine and guanosine mononucleotides, two phases were observed in the binding process: at low values of r (number of bound nucleotides per amino acid residue) the binding isotherm can be fitted by a Langmuir isotherm indicating no interaction between binding sites; at high values of r, the binding is cooperative and accompanied by phase separation or by precipitation of the polypeptide–nucleotide complex (Wagner, 1969); Woese, 1968. Cooperativity in the formation of an ordered poly-L-lysine–GMP complex has also been demonstrated by CD investigations (Wulf *et al.*, 1970). This is accompanied by a change in the conformation of the polypeptide (Rifkind and Eichorn, 1970). Different interactions appear to be involved in these binding processes: electrostatic interaction between the positively charged amino acid and the negatively charged nucleotide; stacking interactions of the bases along the polypeptide; direct interactions between bases and amino acid residues.

II. Structure and Conformation of Polynucleotides

Numerous studies have been devoted to the structures and the conformations of polynucleotides in solution (see reviews: Felsenfeld and Miles, 1967; Michelson *et al.*, 1967; Davies, 1967; Steiner and Millar, 1970; Brahms and Brahms, 1970). We shall mainly deal with polyriboadenylic acid (poly A) and polyribouridylic acid (poly U). This really does not mean that important results have not been obtained with other polynucleotides. For example, Gulik *et al.* (1970) have studied the conformation of polyribocytidylic acid by small angle X-ray scatteiing and spectroscopy. Maurizot *et al.* (1971) have compared oligodesoxyribonucleotides of various chain lengths and oligoribonucleotides in their ability to form complexes. Though the structure of polyriboguanylic acid is not yet known, the conditions of formation of complexes between polyriboguanylic acid and polyribocytidylic acid and the stoichiometry were determined at neutral and acid pH (Pochon and Michelson, 1965; Thiele and Guschlbauer, 1971; Green and Malher, 1970). The interactions between polyribothymidylic acid and polyriboadenylic acid have just been reinvestigated (Howard *et al.*, 1971). The study of polynucleotides containing thiopyrimidine nucleotides are in progress (Scheit and Faerber, 1971; Gottschalk *et al.*, 1971). No attempt in this review will be done to summarise all the results on polyribonucleotides and polydeoxyribonucleotides.

A. SINGLE STRANDED POLYMERS

At pH 7 and room temperature, poly U is thought to be largely devoid of organised structure. The mean results leading to this conclusion are the following (see also a recent review by Steiner and Millar, 1970): X-ray diffraction patterns of fibres are of an amorphous character (Rich, 1957); the hypochromism of poly U at 260 nm is very small and almost independent of temperature (Warner, 1957); the pK of uracil residues in poly U is slightly shifted as compared with UMP (Richards *et al.*, 1963); the optical rotatory dispersion of poly U can be fitted by one term Drude equation (Lipsett, 1960). It must be noticed that Michelson and Monny (1966) have questioned the validity of some of the conclusions drawn from these experimental results. On the other hand, the hydrodynamic studies performed on unfractionated samples are characteristic of a normal flexible chain polyelectrolyte (Richards *et al.*, 1963).

The picture of poly U is an extended coil with small or no stacking interaction at room temperature and neutral pH.

Alkaline poly A is also a single stranded form, but there is good evidence for the presence of some form of organisation (Steiner and Millar, 1970; Witz and Luzzati, 1965). A characteristic temperature dependent transition has been studied in detail by absorption spectroscopy, CD and ORD (Holcomb and Tinoco, 1965; Leng and Felsenfeld, 1966; Brahms *et al.*, 1966; Poland *et al.*, 1966; Applequist and Damle, 1966). It is generally believed that the temperature dependent optical properties arise from the tendency of the bases of poly A to stack. The extent of stacking is temperature dependent, but the stacked structure is formed by a non-cooperative process.

It was of considerable interest to know more on these random coils. In principle, hydrodynamic techniques (viscosity, sedimentation, light scattering) are powerful tools to determine the dimensions of a polymer. However, a complete analysis is quite often hampered by the polydispersity of the sample and the quality of the solvent.

In the simple case of a polydisperse homopolymer which can be characterised by one parameter λ, the different average values of (number average, weight average, *z*-average . . .) can be calculated when the distribution functions are known. In general, the distribution functions are unknown and the interpretation of the experimental results is difficult. For example, light scattering yields the weight average molecular weight and the *z*-average radius of gyration for a statistical coil.

Dimensions of a polymer are dependent on the quality of the solvent. Basically, the spatial conformation depends on the short-range interactions and the long-range interactions. The short-range interactions deal with the bond structure and the local interactions between atoms which are near neighbours in sequence along the chain. The long-range interactions involve segment pairs that are not close neighbours along the chain contour, but can come close and interact. The long-range interactions depend not only on the volume of the chains segment, but also on its interaction with the solvent. In a good solvent, the dimensions are much larger than the dimensions of the chain in absence of long-range interactions (referred as unperturbed dimensions) while, in a poor solvent, the dimensions are smaller than the unperturbed dimensions. Flory has shown that at a certain temperature for a given solvent-polymer, the long-range interactions are compensated by the attractive forces. This temperature is called Flory or theta (θ) temperature. At θ temperature, the dimensions of the chain depend only upon the short-range interactions and Gaussian coil statistics can be applied (Flory, 1953, 1969).

The discovery by Eisenberg and Felsenfeld (1967) that θ conditions for polynucleotides exist in the range of temperature 0–100°C has been a decisive step in the study of these single-stranded polymers.

θ-Temperatures were first established with poly A (Eisenberg and Felsenfeld, 1967). It had been noticed that the solubility of poly A was dependent upon temperature. In high salt concentration (above 1 M NaCl) poly A is soluble at low temperature, precipitates at about room temperature and is again soluble at high temperature. The temperatures of precipitation vary with the molecular weight. This was used to fractionnate poly A. Then the θ-temperatures were determined from classical phase separation studies on the different fractions (Flory, 1953). To each salt concentration, two θ temperatures were found.

θ-Temperatures can also be deduced from light scattering measurements. The theory predicts that the second virial coefficient A_2 vanishes under θ conditions. This was found experimentally and the values obtained from phase separation experiments were confirmed.

When the dimensions of the polymer are not small as compared to the wavelength of the incident light, the study of angular dependence of the scattered light permits the determination of the radius of gyration. The radii of gyration of poly A fractions were found temperature dependent, under θ conditions.

For each fraction, sedimentation coefficient distributions and intrinsic viscosity $[\eta]$ were determined. The results were fitted to expression of the form

$$s = K_s Z^{a_s}$$

$$[\eta] = K_n Z^{a_n}$$

K_s, a_s, K_n, a_n are constants, Z is the degree of polymerisation. The polydispersity of the sample has been taken into account and consequently these relationships are related to homogeneous samples (the molecular weight distribution was calculated by fitting of the sedimentation coefficient distributions to the molecular weights determined by light scattering). It was found $a_s = a_n$ at θ-temperature as it is predicted by the theory.

Similar study was performed on poly U (Inners and Felsenfeld, 1970). Fractions were obtained by precipitation of poly U in hydroalcoholic solutions. Theta temperatures were deduced from light scattering and confirmed by sedimentation and viscosity measurements. Unperturbed dimensions of poly U were found to vary slightly with temperature, as expected for a polymer with little or no stacking.

Many important results were obtained from these studies on poly A and poly U as the constants P and ϕ ($f = P\eta_0\langle h^2\rangle^{1/2}$, $[\eta] = \phi(\langle h^2\rangle^{3/2}/\mathrm{M})$, where f is the frictional coefficient of the polymer, η_0 the viscosity of the solvent, $\langle h^2\rangle$ the mean square end-to-end distance of the coil). However we shall focus our attention on the following points:

The unperturbed dimensions of poly U are those of an extended coil.

The unperturbed dimensions of poly A are function of temperature.

A convenient quantity to characterise a random coil is the ratio $C_n = \langle h^2\rangle/nl^2$, n being the total number of bonds in the backbone, l^2 the mean square bond length (equal to 2·31 Å in poly A or poly U). C_n is equal to unity in a freely jointed chain and becomes larger as the hindrance to free rotation increases. The value of C_n is about 7 for polyphosphates (Strauss and Ander, 1962) and is found to be 17–18 in the case of poly U. This means that poly U is a rather extended coil and that the short-range interactions are important.

The causes of this high value of C_n are not yet known. An hydrogen bond between the 2′–OH and an atom of the base or the backbone has been postulated (Ts'o *et al.*, 1966; Brahms and Sadron, 1966). This could contribute to the rigidity of the chain. However, in a recent study of the conformation of apurinic acid (single-strand polydeoxyribonucleotide in which almost half of the bases has been removed) Atcher and Felsenfeld (1971) found that C_n was equal to 18. This hydrogen bond, if it exists in polyribonucleotides, is not a major factor for the stiffness of the chain. Another possibility would be that the electrostatic interactions between neighbouring phosphates are still important, despite the fact the experiments are performed in high salt concentration.

TABLE III

Models of torsional angle restriction in polynucleotides Taken from Inners and Felsenfeld (1970)

Angle†	Between	Model 1‡ (°)	Model 2 (°)	Model 3 (°)	Model 4§ (°)
$\omega'(\phi)$	C3′O3′P and O3′PO5′	190–280	270–290	190–280	171–280
$\omega(\psi)$	O3′PO5′ and PO5′C5′	257–315	281–315	257–315	257–315
$\phi(\theta_1)$	PO5′C5 and O5′C5′C4	140–210	140–210	130–250	130–250
$\psi(\theta_2)$	O5′C5′C4′ and C5′C4′C3′	50 and 175	50 and 175	60 and 175	60 and 175
ψ'	C5′C4′C3′ and C4′C3′O3′	83 and 150	83 and 150	83 and 150	80 and 150
$\phi'(\theta_3)$	C4′C3′O3′ and C3′O3′P	185–244	185–244	185–244	185–290
$\langle h^2\rangle/nl^2$		17·5	13·1	14·9	11·2

† The names in this column are those given by Sundaralingam (1969). The names used by Sasisekharan and Lakshminarayanan (1969) are shown in parentheses. In order to avoid confusion we follow these workers in assigning zero angle to the *cis* configuration about each bond.

Using the convention of Flory (1969) for the numbering of bonds and interbond angles, the bond distances corresponding to the dihedral angles, starting from the top of the table, are 1·6, 1·6, 1·47, 1·514, 1·523 and 1·404 Å. The complements of the bond angles are 76, 62, 71, 63·9, 64·7 and 62 degrees, respectively.

‡ A dash between angles indicates that an infinitely deep square well potential was assumed, with its walls located at the limits shown. The word "and" between angles indicates that these two fixed angles were used, but the angles in between were excluded.

§ Model 4 represents what we consider to be the extreme of flexibility derivable from Sundaralingam's results. In all of these models we have excluded the values ($\omega=60°$, $\omega'=60°$), which are a coupled pair observed only in diethyl phosphate, the protonated form of 3′ nucleotides, and glycerylphosphorylcholine.

Whatever are the exact causes, the relative rigidity of the coil indicates that the rotation of the bonds of polynucleotide backbone is restricted. This has been already proposed from steric considerations (Sasisekharan *et al.*, 1967). More recently, Sundaralingam (1969) and Arnott (1970), from X-ray diffraction of nucleotides and polynucleotides concluded that the torsional angles of the backbone lie within characteristic limits. Also, from the study of the ^{13}C magnetic resonance spectra of poly U, uridine and related nucleotides it is found that the polynucleotide backbone has distinct conformational preferences (Mantsch and Smith, 1972). Inners and Felsenfeld (1970) have chosen four sets of values for the torsional angles (among those given by Sundaralingan, 1969) and computed C_n using the method of Flory and Jernigan (1965). The calculated values are comprised between 17·5 and 13·1 (Table III).

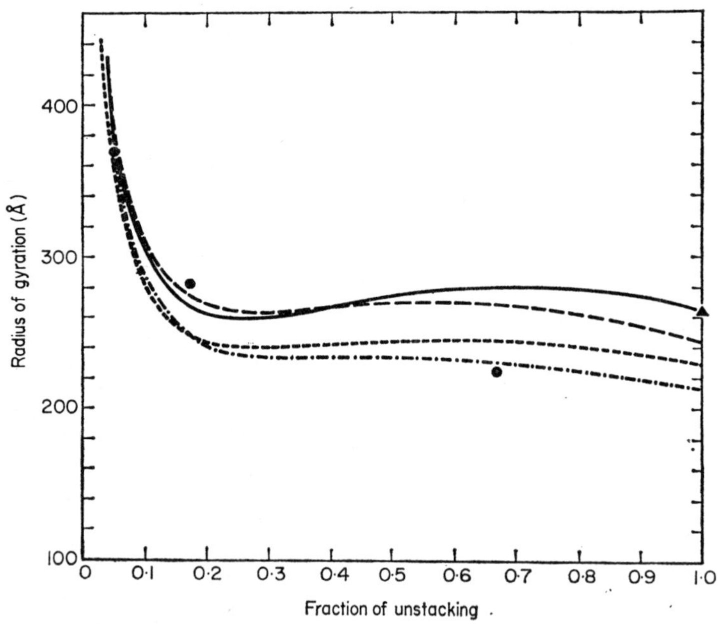

FIG. 3. Temperature dependence of the radius of gyration of poly A. The dependence of radius of gyration upon fraction of bases unstacked, calculated for a hypothetical chain containing 1740 nucleotides for four possible sets of restriction on torsional angles in the unstacked form (see Table III). (————) model 1; (— — — —) model 2; (— — — —) model 3; (—·—·—·) model 4. (●) Poly A experimental values (Eisenberg and Felsenfeld, 1967) of unperturbed dimensions of fraction with a degree of polymerisation 1740 corrected for heterogeneity; (▲) value of unperturbed radius of gyration to be expected for a homogeneous fraction of poly U with a degree of polymerisation 1740. (Taken from Inners and Felsenfeld, 1970.)

The next step was to compute the dimensions of the coil upon the extent of stacking (Inners and Felsenfeld, 1970). This corresponds to the case of poly A. It was assumed that the stacked form has the torsional angles of one strand of RNA–10, while the torsional angles of the unstacked form correspond to one of the four sets of values given in the Table III. The values of the radii of gyration of a chain of DP=1740 as a function of the fraction of unstacking are reported on Fig. 3. This shows that the unperturbed dimensions of unstacked poly A are slightly smaller than that of poly U. In fact, the dimensions of unstacked forms are probably independent of the nature of the bases. The calculated values might be subject to small changes since different assumptions have been made. It was assumed that the denaturation is non-cooperative, and that the torsional angles of the back bone in solution are the same as those deduced from X-ray diffractions on fibres.

B. MULTI-STRANDED STRUCTURES

1. *Homopolymers*

Poly A and poly U can form ordered structures with hydrogen bonds between the bases. By lowering the pH below about pH 6, poly A undergoes a sharp structural transition. The ordered form consists of two helical strands, the strands being parallel. The thermal stability of acid poly A depends upon ionic strength and pH. The main results have been summarised (Steiner and Millar, 1970).

At neutral pH, Lipsett (1960) has observed that on cooling a solution of poly U, a sharp structural transition also occurs. This was often confirmed, but nothing precisely was known on this structure. Recently, Thrierr and Leng (1969) found that the molecular weight was unchanged during the transition. This fact precluded the possible association of different strands of poly U. A model in which poly U folds on itself was deduced from the following results (Dourlent *et al.*, 1970; Thrierr *et al.*, 1971).

i. There is no variation of the molecular weight in a large range of temperature above and below T_m.

ii. The intrinsic viscosity decreases as the temperature decreases, goes through a minimum, then increases. Moreover, while the intrinsic viscosity of the disorder form is larger at low salt concentration than at high salt concentration, the intrinsic viscosity of the ordered form is almost independent of the salt concentration.

iii. Thermal stability also depends on ionic strength, but the transition becomes broader as the salt concentration increases. No dependence of T_m with the concentration of oligo U or of poly U was observed (in the range 10^{-4}–10^{-3} M).

iv. Temperature-jump experiments show that the relaxation curves are strongly dependent on the final temperature T_f reached after the equilibrium perturbation. Three typical regions can be distinguished. For $T_f < 9°6$ ($T_m = 10°$ in 0·5 M CsCl) there is a distribution of relaxation times. For $T_f < 12°$, several relaxations times are present, but it is possible to measure the longer one (its value is between 0·13 and 1·0 msec). For T_f higher than 14°, almost the whole signal involves only one relaxation time of the order of 100 μsec.

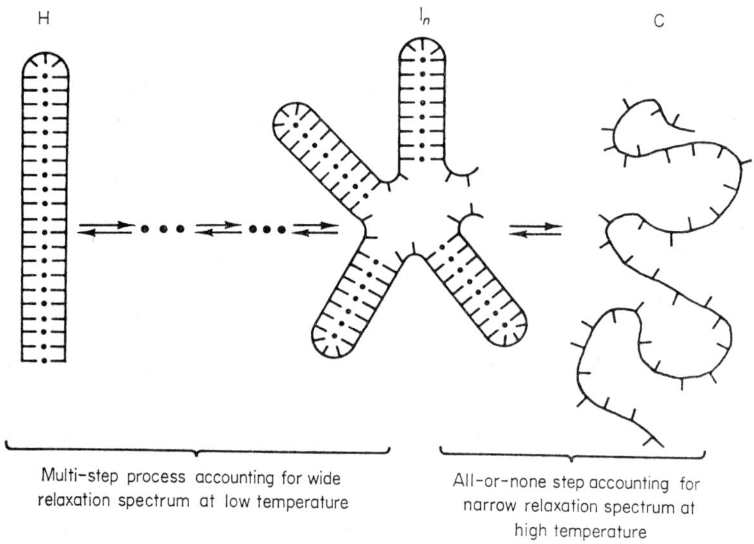

FIG. 4. Schematic representation of poly U structural transition. (Taken from Thrierr, Dourlent and Leng, 1971.)

These experimental results led to the proposal of the model represented on Fig. 4. Another model assuming a single-strand polymer in which the stacking increases as the temperature is decreased is not consistent with the viscosity data. Moreover, there are good evidence of hydrogen-bonds between bases in the structured poly U (Lewin, 1964; Aylward, 1966). Also, poly N_3-methyl U (Szer and Shugar, 1963), poly U after reaction with methyl mercury hydroxide (Millar, 1970) and poly U at basic pH (above pH 10) do not form the helical structure. This implies that the proton on the N_3 is involved in the hydrogen-bonding. However, the exact base-pairing is not yet established.

Folding of poly U also occurs in other monovalent salts, but salt dependent aggregation was observed (Brown, 1966; Millar and MacKenzie, 1970). On the other hand, in salt-free aqueous methanolic solvent, at low temperature, the structured form is assumed to result from the interaction between separate strands, the strands being parallel (Hui Bon Hoa *et al.*, 1969).

A sharp structural transition was observed for poly U in presence of diamines, spermidine, spermine (Szer, 1970). In water, the hyperchromicity of the complex poly U–spermine (4/1 in mol) is larger than that of poly U alone in salt solution and chiefly the T_m is very high (about 30°). ORD spectra of these complexes (Rogers *et al.*, 1967) have been reported, but the exact structure is still unknown.

Systematic study of polynucleotides of the ribose, 2′–O methylribose and deoxyribose of uracil and thymine will certainly help to elucidate the importance of the different forces involved in these structures (Zmudzka and Schugar, 1971; Singh and Hillier, 1971).

2. *Interactions between Poly A and Poly U*

Two well-defined complexes can be formed on mixing poly A and poly U. In one of them, the monomer molar ratio of adenine to uracil is 1/1 and the structure is a two-stranded helix. In the other one, the ratio is 1/2 and the structure is a three-stranded helix. The stability of one or the other is determined by salt concentration, temperature, pH. At neutral pH a "phase diagram" has been established (Stevens and Felsenfeld, 1964; Massoulie, 1968).

Poly A.poly U and poly A.2 poly U are very often used as model compounds. However, the structure of these products is not yet known with certainty. Poly A.poly U was believed to have a structure similar to that of DNA, but Sasisekharan and Sigler (1965) found that the diffraction pattern of this complex resembled that of double-stranded helical RNA (Arnott, 1970).

Two different structures have been assumed for poly A.2 poly U. The original structural proposal (Felsenfeld *et al.*, 1957), based on model building studies is represented on Fig. 5. The two strands of poly U are anti-parallel. On the other hand, i.r. studies have led Miles (1964) to question this model. He has proposed a different one, in which the two strands of poly U are parallel.

No systematic work on the hydrodynamic properties of poly A.2 poly U has yet been done. Fluorescence polarisation measurements (poly U is conjugated with label acriflavine) have indicated that poly A.2 poly U is more rigid than poly A.poly U (Millar and Steiner, 1965).

Recently, Thrierr and Leng (1972) have studied the conversion 2 poly A.poly U$\rightleftharpoons$poly A.2 poly U+poly A by light scattering. The experiments were performed on large molecular weight polymers in high salt concentration (1·2–1·5 M NaCl). The question was to know whether, above the T_m of the conversion, there was completely free macromolecule of poly A or,

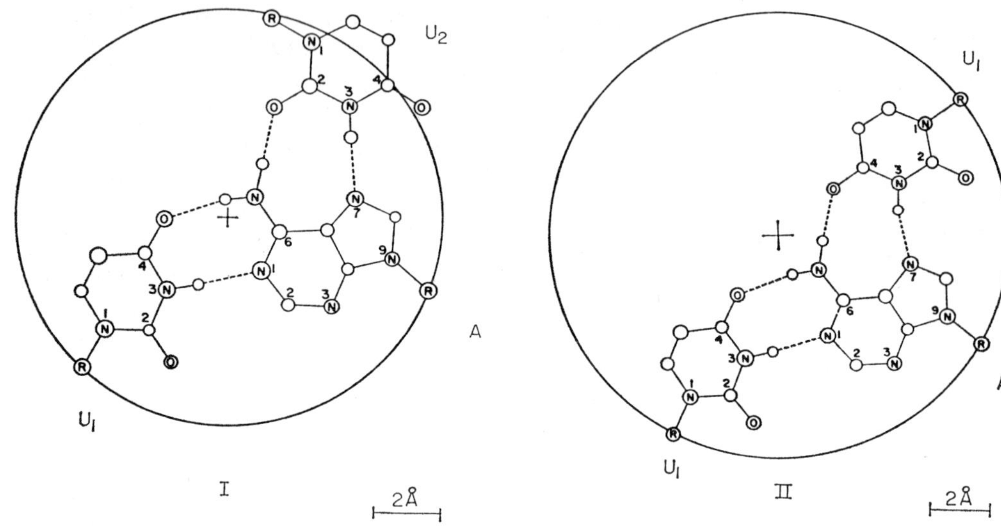

FIG. 5. Possible nature of hydrogen bonds between poly A and poly U. (Miles, H. T. (1964). *Proc. natn. Acad. Sci. U.S.A.* **51**, 1196.)

every poly A macromolecule was complexed with poly U, a part of poly A interacting with 2 poly U, the rest of the molecule being free of poly U. Experimentally it was found that when the molecular weight of poly A is larger or equal to that of poly U, at the temperature of the conversion and above, there was no free poly A. Therefore, the conversion occurred according to an intramolecular process. When the molecular weight of poly A was smaller than that of poly U, free poly A was detected. From the analysis of the molecular weights these results were interpreted assuming a folding of poly U on poly A whatever are the relative sizes of poly A and poly U.

On the other hand, the molecular weight of poly A.2 poly U prepared by direct mixing of a large poly U and a small poly A (mol ratio 2/1) was measured. The molecular weight of the complex is smaller than three times the molecular weight of poly U, but slightly larger than one and half the molecular weight of poly U. Folding of poly U on poly A was also assumed in this case. In this model, the two strands of poly U have to be anti-parallel and therefore the atoms involved in hydrogen bonds are those postulated by Felsenfeld *et al.* (1957). However, some restrictions have to be done on the folding of poly U on poly A. It has been shown that poly A can interact with UpU and three-stranded structure is formed (Felsenfeld and Miles, 1967). A three-stranded structure is also made on mixing oligo A and oligo U (pentanucleotides), but one has to work at relatively high concentrations of oligonucleotides (Cantor and Chin, 1968). Folding of UpU or oligo U in these complexes is quite unlikely, but the U strands can still be anti-parallel.

Finally, it has to be mentioned that the complexes between poly A and poly U and chiefly, between oligo A and oligo U are simple and convenient models to investigate the dynamics of the process of base pairing (Pörschke and Eigen, 1971, Pörschke, 1971).

III. Interactions between Nucleic Acids and Small Aromatic Molecules

A. DYE BINDING ON DOUBLE-STRANDED STRUCTURES

Dye molecules (e.g. proflavine, acridine orange, ethidium bromide) can be considered as model compounds for the study of the general features of the interactions between polynucleotides, nucleic acids and small molecules because these dyes have structural similarities with a number of other molecules of biological interest. These similarities are the presence of flat aromatic rings and a positive charge at neutral pH (Fig. 6). Moreover, aminoacridines and ethidium cation can absorb and emit light in the visible region where biological macromolecules are transparent. This property allows to excite selectively the dye molecule thus facilitating the interpretation of experimental data in optical studies.

In the following sections no attempt will be made to give a complete report on the very wide and complex subject of dye interaction with nucleic acids or polynucleotides (for a more detailed description see the survey of Blake and Peacocke, 1968). Regarding these dyes as model compounds we

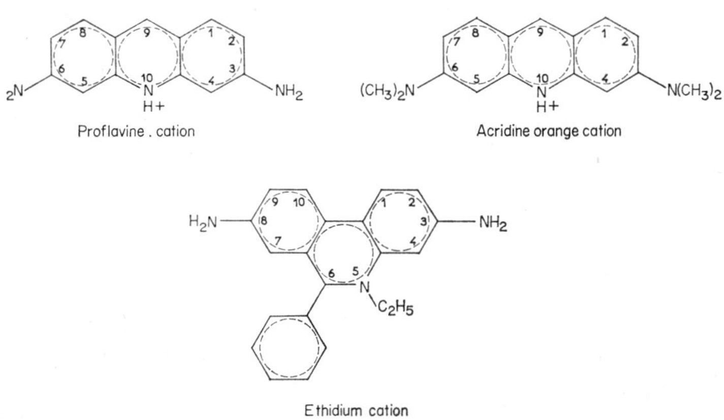

FIG. 6. Formulae of proflavine, acridine orange and ethidium cations.

will rather focus our attention on the mechanism of binding and on the structural modifications of the polymer induced by ligand binding.

It is now well established that the interaction of aminoacridines and ethidium bromide with double helical nucleic acids and polynucleotides involves *at least* two kinds of complexes. The existence of a third complex must be considered on the basis of recent kinetic data (see Section IIIA, 3.). For all complexes, the extent of binding decreases as the ionic strength increases showing that electrostatic forces play a role in the interactions. For the weaker binding process the influence of the ionic strength is more important than for other ones so that it is possible to choose experimental conditions ($\Gamma \geqslant 0{\cdot}2$ M monovalent salt) allowing to obtain a negligible amount of weakly bound dye over a wide range of phosphate (P) to dye (D) ratios.

1. *Equilibrium Data*

Under conditions allowing to neglect the weaker binding process the experimental data always show that the number r of bound dye molecules per phosphate never exceed 0·25 (Bauer and Vinograd, 1970; Armstrong *et al.*, 1970). Plotting r/c versus r (Scatchard representation, where c is the concentration of unbound dye) the binding isotherm exhibits a curvature even in the absence of weakly bound dye (e.g. Bauer and Vinograd, 1970—Fig. 1). It might be thought that the limitation in the extent of binding and the non-linearity of the Scatchard plot could be due to an *a priori* limited number of potential binding sites or a non-equivalence between them. Indeed, it has been proposed that A.T. pairs have a larger affinity for the dye than G.C. pairs (Tubbs, *et al.*, 1964; Thomes *et al.*, 1969; Bidet *et al.*, 1971) and that the sites of highest binding affinity are those situated between adjacent thymines in a single strand of DNA (Roth and Manjon, 1969). However these conclusions are challenged (Chan and Van Winkle, 1969; Ellerton and Isenberg, 1969) and the limitation in the extent of strong binding is also observed with double helical poly A.poly U for which all base pairs are identical (Schmechel and Crothers, 1971). A model consistent with these findings is the nearest neighbours exclusion model (Bauer and Vinograd, 1970; Armstrong *et al.*, 1970): one potential binding site is assigned to each base-pair with the restriction that binding cannot occur at all to the two sites immediately adjacent to the occupied one. This model belongs to the general class of nearest neighbour cooperative binding processes† which have been described

† The nearest neighbour cooperative model considers two different binding constants: the nucleation binding constant (K) which represents the binding constant to a site whose nearest neighbours are unoccupied and the growth binding constant (K') which represents the binding constant to a site adjacent to an occupied one. The ratio $q=K'/K$ measures the degree of cooperativity: $q>1$ corresponds to positive cooperativity; $q<1$ corresponds to negative cooperativity (or anti-cooperativity). The exclusion model corresponds to $q=0$.

by Schwarz (1970). The exclusion model corresponds to the special case of strongest negative cooperativity (or anti-cooperativity). In the present state of accuracy of experimental determinations, models displaying less strong anti-cooperativity could also account for the experimental data relative to dye binding on double stranded nucleic acids.

The weaker binding complex becomes apparent at small ionic strength and low P/D values but it cannot be observed alone with double helical polymers because of the competitive strong binding process. The weak binding is mainly electrostatic (as shown by its strong dependence on the ionic strength) and also involves interactions between bound dye molecules as shown by the spectral shift which is characteristic of the formation of dye aggregates along the polymer chain (Blake and Peacocke, 1968; Stone *et al.*, 1970). Using very low salt concentration (10^{-3} M Na cacodylate buffer) and rather large proflavine concentration ($9 \cdot 10^{-5}$ M) Stone *et al.* (1970) have shown a one-to-one dye per phosphate stoechiometry. A close correlation exists between the aggregation tendency on nucleic acid and the tendency to self aggregation in pure dye solutions: acridine orange has a strong tendency to form aggregates, this tendency is less strong with proflavine and only weak with ethidium bromide (but still existing as shown by recent studies from Kreishman *et al.*, 1971). The weak binding on DNA has many features similar to those observed in dye binding on other polyanions as acidic polysaccharides (Stone and Bradley, 1967; Stone, 1967) poly L-glutamic acid (Schwarz *et al.*, 1970; Schwarz and Balthasar, 1970) polyphosphates (Weill and Calvin, 1963) and single-stranded polynucleotides (Dourlent and Hélène, 1971). A theoretical model accounting for this type of binding is the nearest neighbour cooperative model described by Schwarz (1970). The constancy of the similarities observed with a number of different polyanions suggests that electrostatic interactions between the dye and the phosphate groups of nucleic acids and cooperative interactions between neighbouring bound molecules are the major sources of stabilisation for the weak complex although a possible contribution of the bases cannot be completely excluded.

2. *Structural Features*

It is generally agreed that, in the strong complex, the dye molecule is intercalated between two consecutive base-pairs, the planar aromatic rings being parallel to the base planes. Although no direct proof of this structure has yet been obtained, there are a number of indirect evidences supporting this idea (Blake and Peacocke, 1968; Fuller and Waring, 1964; Neville and Davies, 1966), the strongest one being the fact that the contour length of rodlike low molecular weight DNA is increased on binding as shown by viscosity and sedimentation studies (Cohen and Eisenberg, 1969).

The first molecular model for aminoacridine intercalation was proposed by Lerman (1961): the aminoacridine cation lies centrally between two successive base pairs so that the positive nitrogen atom is near the central axis of the DNA molecule and equidistant to the two polynucleotide chains. In order to accommodate the dye the double helix was assumed to locally untwist and extend so that the distance between the base-pairs between which the dye is sandwiched is approximately twice their distance in the absence of bound molecule. Experimental determinations of the contour length increase (Cohen and Eisenberg, 1969) agree with this value. A modified model has been proposed (Pritchard *et al.*, 1966) in which the acridine cation lies between successive nucleotide bases on the same polynucleotide chain in a plane approximately parallel to the base planes but at an angle (looking down on the polynucleotide chain) such that the positive ring nitrogen is close to the polymer phosphate group. One of the arguments for this proposal was that strong binding occurs to about the same extent in native and denatured DNA (Drummond *et al.*, 1965) but without increase in contour length in this latter case. Recent studies on single stranded polynucleotides (see section IIIB) can also argue for this model. However, it must be pointed out that the structure of the complex could be very different in single stranded and double stranded polymers.

In the case of ethidium cation, Fuller and Waring (1964) have proposed a model similar to that of Lerman for DNA-proflavine complexes with, however, a smaller untwisting angle of the polymer (12° instead of 45°). On the basis of this proposal, the studies relative to ethidium binding on closed circular DNA lead to conclude that the tertiary turns on the macromolecule are negative (Vinograd *et al.*, 1968; Bauer and Vinograd, 1970; Crawford and Waring, 1967a, b; Bujard, 1968; Waring, 1970). However an opposite conclusion is drawn by Paoletti and Le Pecq (1971a) from their proposal that the DNA helix is not unwound but wound to an angle of $13° \pm 4°$ after ethidium intercalation. This value results from a comparison between experimental data on fluorescence depolarisation and computer simulation of energy transfer between bound ethidium cation. A modified molecular model has also been proposed (Paoletti and Le Pecq, 1971a) which differs from that of Fuller and Waring (1964) but has some similarity with that of Pritchard *et al.* for DNA-acridine complexes: in the model of Paoletti and Le Pecq the negative charge of the phosphate group interacts with the positive charge of the ethidium cation which has been shown to be on the nitrogen in position 8 in crystals (Hospital and Busetta, 1969). Winding of the DNA helix can be obtained only if one assumes that distorsion of the sugar-phosphate backbone occurs not only at the intercalation site but also at the two nearest neighbour sites (Paoletti and Le Pecq, 1971b) thus providing a molecular basis for the nearest neighbour exclusion model discussed in Section IIIA, 1.

Experiments performed on other drugs have shown that the torsion angle depends on the nature of the intercalating compound (Saucier *et al.*, 1971); if the torsion angle is taken 12° for ethidium cation, the following values are obtained; 8–9° for proflavine and quinacrine, 7° for methoxyellipticine and only 4° for daunomycine.

As regarding the structural features of the weak complex, a model has been proposed by Bradley and co-workers (Bradley and Wolf, 1959; Stone and Bradley, 1961). In this model the dye molecules are bound externally to the polyelectrolyte chain by electrostatic forces. Aggregated bound dye molecules are stacked on top of each other in the direction of the helix axis with their planes parallel to the base planes and their positive charge close to the external phosphate groups.

On the basis of equilibrium dialysis experiments at low ionic strength (2×10^{-3} M) Armstrong *et al.* (1970) proposed that for $r \leqslant 0{\cdot}2$ two spectroscopically distinct species of bound dye exist: one of these species is the intercalated monomeric dye, the other one is a bound dimer formed by addition of a non-intercalated dye molecule to an intercalated one. If the additional bound molecule corresponds to the monomeric form of the weak complex this could mean that the nucleation sites for the cooperative binding are preferentially at those sites where and intercalated dye molecule is present.

3. *Kinetic Data*

Temperature-jump relaxation and stopped flow techniques have been used to study dye binding to DNA, poly A.poly U and transfer RNA (Bittman, 1969; Li and Crothers, 1969; Schmechel and Crothers, 1971; Akasaka *et al.*, 1970; Sakoda *et al.*, 1971; Tritton and Mohr, 1971). In the strong binding region, at low *r* values, the results are consistent with a two-step mechanism first proposed by Li and Crothers (1969).

$$P+\text{DNA} \underset{k_{21}}{\overset{k_{12}}{\rightleftharpoons}} P_{(\text{out})} \underset{k_{32}}{\overset{k_{23}}{\rightleftharpoons}} P_{(\text{in})}$$

In this scheme P represents the free dye, DNA is an unoccupied binding site, $P_{(\text{out})}$ and $P_{(\text{in})}$ are two complexes respectively called "outside-bound" and "intercalated" forms of the dye. In the case of proflavine, all rate constants decrease as the ionic strength increases (between 0·016 and 0·2 M) showing again the contribution of electrostatic forces in the intercalation process. However an opposite variation of the intercalation rate has been found in the case of acridine orange (Sakoda *et al.*, 1971). Values of rate and equilibrium constants are summarised in Table IV (note that the standard reaction enthalpies and entropies are much dependent on the nature of the polymer).

The nature and the structural features of the so called "outside-bound" form are still obscure. Some of its properties argue for intercalation, others

TABLE IV

Thermodynamic and kinetic parameters for the intercalation in double stranded structure according to the mechanism of Li and Crothers (1969)

Dye	Proflavine					Ethidium bromide
Macromolecule	Calf thymus DNA†			T 2 DNA†	$A_n.U_n$‡	Transfer§ RNA
Temperature (°C)	10	10	10	10	10	25
Ionic strength (M)	0·016	0·2	1·0	0·2	0·2	0·2
pH	7·1	6·9	6·5	6·9	6·9	7·0
k_{12}($M^{-1} \times s^{-1}$)	—	$2·1 \times 10^7$	—	—	$4·3 \times 10^6$	$5·67 \times 10^5$
k_{21} (s^{-1})	—	$3·5 \times 10^3$	—	—	$4·8 \times 10^2$	$8·66 \times 10^2$
k_{23} (s^{-1})	$1·2 \times 10^4$	$1·5 \times 10^3$	$2·2 \times 10^3$	$2·5 \times 10^2$	64·8	$4·68 \times 10^2$
k_{32} (s^{-1})	$5·0 \times 10^3$	$1·1 \times 10^2$	$1·1 \times 10^2$	40	8·28	$6·23 \times 10^2$
K_{12} (M^{-1})	$6·8 \times 10^4$	$5·9 \times 10^3$	$6·2 \times 10^2$	$1·3 \times 10^4$	$8·25 \times 10^3$	$6·55 \times 10^2$
K_{23}	2·4	13·6	20	6·2	7·8	0·75
$\Delta H_{12}°$ (kcal/mol)		−9·8		−3·4	−1·0	
$\Delta H_{23}°$ (kcal/mol)		2·0		−3·9	−7·6	
$T.\Delta S_{12}°$ (kcal/mol)		−5·4		1·6	3·6	
$T.\Delta S_{23}°$ (kcal/mol)		3·4		−2·8	−6·6	

For DNA and poly A.poly U the polymer concentration is expressed in base-pair equivalent per litre.

For *t*RNA the polymer concentration is expressed in phosphate equivalent per litre.

† Li and Crothers (1969).

‡ Schmechel and Crothers (1971).

§ Tritton and Mohr (1971).

against. In the case of proflavine, the absorption spectrum of $P_{(out)}$ has been found similar to that of $P_{(in)}$ (Li and Crothers, 1969) and this indicates some coupling between the base pairs and the aromatic rings of the "outside-bound" dye molecule. On the other hand, viscosity and sedimentation data (Schmechel and Crothers, 1971; Cohen and Eisenberg, 1969) are in excellent agreement with a model in which $P_{(out)}$ does not contribute to the increase of the molecular length of rodlike DNA and poly A.poly U. However it seems very likely that $P_{(out)}$ does not correspond to the weaker complex because at the r values and ionic strength conditions used in the kinetic studies of Li and Crothers (1969) the amount of weakly bound dye was actually quite negligible (Armstrong *et al.*, 1970). One is therefore faced with the following alternative: is the dye molecule able to interact with the DNA bases without any intercalation in the helix, or, is the dye molecule able to intercalate (at least partially) in the DNA helix without significantly increasing its contour length?

In conclusion, kinetic studies reveal that the actual mechanism for dye binding in the strong binding region is not as simple as predicted by models derived from equilibrium studies and that a complex preceding the deeper intercalated form must be considered. However, in medium and large ionic strength conditions ($\geqslant 0{\cdot}2$ M) the amount of "outside-bound" molecules is small and do not alter significantly the conclusions that can be drawn from equilibrium studies (at least for proflavine-DNA complexes).

B. DYE BINDING ON SINGLE-STRANDED POLYMÈRS

Recent studies on proflavine–poly A and proflavine–poly U systems (Dourlent and Hélène, 1971) have shown that the interaction can be described by the consideration of two binding processes similar to those found is double helical molecules: electrostatic cooperative binding along the phosphate backbone and further binding resulting from interactions with the bases. This analysis rests on the fact that the total absorption spectrum can be resolved in three components spectra respectively assigned to: electrostatically bound monomer dye (absorbing as free molecule), stacked electrostatically bound dye (absorbing as stacked aggregates bound to polyanions) and "intercalated" dye (absorbing as intercalated dye in DNA). The cooperative model is in good agreement with the electrostatic binding process and the anti-cooperative model can also account for the other binding process but with a large indetermination in the degree of anti-cooperativity (a non-cooperative model could also account for the results).

Although it cannot be concluded from these studies that proflavine does actually intercalate between successive bases in single-stranded polynucleotides, this can reasonably be thought from the p.m.r. data of Kreishman *et al.* (1971) on the interaction of ethidium bromide with several uracil residues, uridyl 3′–5′ uridine and poly U. In these studies more convincing evidences for the formation of an intercalation complex are obtained from the upfield chemical shifts of UpU resonances induced on ethidium bromide binding. A model has been proposed in which the ethidium cation is sandwiched between the adjacent uracil bases and oriented with the H (1) and H (10) protons directed at the ribose phosphate backbone while the quaternary nitrogen and phenyl ring are on outside in contact with the solvent (Kreishman *et al.*, 1971).

If confirmed by further work, the results obtained on single-stranded polynucleotides show that intercalation can occur even on coiled chains, as shown by the binding to poly U. However, the evaluation of the relative binding strengths show that, in this case, the nucleation binding constant for intercalation is smaller than the nucleation constant for electrostatic binding. Therefore the amount of intercalated dye is always less than the amount of externally bound dye. In contrast, the results obtained on poly A and double-stranded poly U show that as the polymer is in a stacked conformation, the nucleation constant for intercalation becomes larger than the nucleation constant for electrostatic binding (Dourlent and Hélène, 1971).

On a general standpoint (including DNA) and using the nearest neighbour cooperative model (Schwarz, 1970) the competition between the two binding processes depends on two pairs of thermodynamic parameters q and K. Let $q_c (q_c > 1)$ and K_c stand for the external cooperative binding, and $q_i (q_i \leqslant 1)$ and K_i stand for the intercalation process. The theory of cooperative binding (Schwarz, 1970) shows that at small P/D values the external binding is mainly governed by the growth constant $q_c K_c$ while at large P/D values the significant term is the nucleation constant K_c. If $q_i < 0{\cdot}1$ the anti-cooperative binding mainly depends on K_i. Therefore at the larger P/D values, the ratio of intercalated and externally bound dye approaches K_i/K_c. If $K_c > K_i$ electrostatic binding always predominates. If $K_c < K_i < q_c K_c$ cooperative binding predominates at low P/D values and large free dye concentration, but the relative amount of intercalated dye continuously increases as P/D increases and prevails against the amount of externally bound dye at the larger P/D values. If $K_c < q_c K_c < K_i$, the behaviour is difficult to describe because it greatly depends on the q_c value. However, if K_i is at least one order of magnitude larger than $q_c K_c$ the relative amount of intercalated dye can be predominant at every P/D values.

In the case of poly A, one has $K_c < K_i < q_c K_c$, and it could be shown that, if the sites for external binding and intercalation were non-interacting one should find intercalated molecules at all P/D values in low ionic strength conditions (Dourlent and Hélène, 1971). The experimental data show that at low P/D ratios ($P/D \leqslant 1{\cdot}2$) all the bound molecules are in the external complex, thus suggesting that a given nucleotide can bind a dye molecule either in the electrostatically bound form (state I) or in the intercalated form (state II). The most general mechanism consistent with this idea is:

$$\text{Dye} + \text{free site} \begin{array}{l} \rightleftharpoons \text{occupied site in state I } (P_{\text{I}}). \\ \qquad\qquad\qquad\updownarrow \\ \rightleftharpoons \text{occupied site in state II } (P_{\text{II}}). \end{array}$$

(This scheme is only a qualitative picture because quantitative treatment should take into account the respective cooperativities displayed by the two binding processes.)

Temperature-jump relaxation experiments should help to obtain informations concerning the validity of this model. The present data (Hammes and Hubbard, 1966) do not permit to answer this question because of an unresolved broad relaxation spectrum originating from cooperative external binding (Schwarz, 1970; Schwarz and Balthasar, 1970c).

Due to the fact that electrostatic forces operate at longer ranges than Van der Waals–London forces it can be expected than the above general scheme reduces to a simpler one

$$\text{Dye} + \text{free site} \rightleftharpoons \text{occupied site in state I } (P_{\text{I}}) \rightleftharpoons \text{occupied site in state II } (P_{\text{II}})$$

In such a case it would be tempting to picture the intercalation process as a motion involving some rotation of the electrostatically bound molecule around its positive charge which would constantly remain in interaction with the negative phosphate group, thus favouring the model of Pritchard *et al.* for double-stranded structures.

However, it must be emphasised that this picture is a very simplified description which does not account of a lot of other effects which are specific of either the polymer (see for instance the differences observed between calf thymus DNA and T2 DNA and the influence of salt conditions (Li and Crothers, 1969) or the ligand (see for instance the dependence of the binding on the dye structure studied by Löber and Achtert (1968) which led to the concept of different conformations for the intercalated dye).

C. INTERACTION OF OTHER AROMATIC SMALL MOLECULES WITH NUCLEIC ACIDS

Besides nucleic acids derivatives and dyes, a large number of aromatic molecules of biological interest interact with nucleic acids: purines and pyrimidines derivatives, carcinogenes (polycyclic hydrocarbons and benzacridines) some antibiotics (e.g. actinomycine), amino acids and their derivatives. Purine has been shown to intercalate between bases of single-stranded polynucleotides or denatured DNA by viscosity, sedimentation and ORD (Sander and Ts'o, 1970) as well as p.m.r. measurements (Bangerter and Chan, 1968b). The binding does not result in a more ordered or rigid structure of the polymer.

Polycyclic hydrocarbons are uncharged molecules. Therefore, hydrophobic and stacking interactions play the major role in their interactions with nucleic acids. An intercalation model has been proposed (Boyland and Green, 1962; Liquori *et al.*, 1962) which rests on a set of indirect evidences:

flow dichroism observed on pyrene, 3,4-benzpyrene and phenantrene complexes (Nagata *et al.*, 1966), changes in the polarisation of fluorescence of hydrocarbon complexes with DNA oriented by flow (Green and McCarter, 1967), comparative studies of spectral shifts in different solvents and in DNA complexes (Lesko *et al.*, 1968), and complexes of purines and hydrocarbons in the solid state (Domiani *et al.*, 1965). To test the model, a size criterion has been proposed (Isenberg *et al.*, 1967, 1969) which states that the only hydrocarbons which bind to DNA in appreciable amount will be those that may intercalate and oriente in such a position that they are protected from contact with the aqueous medium. Data obtained on fourteen hydrocarbons (Isenberg *et al.*, 1967, 1969; Craig and Isenberg, 1970a) show that the predictions based on the size criterion are found to be valid with DNA. It was also observed that the binding on double stranded acidic poly A obeys a different size criterion (Craig and Isenberg, 1970b). This is explained by the differences in cross-section between double helical DNA and poly A molecules.

In spite of the absence of electric charge on the small molecule, the apparent binding constant may depend on the ionic strength, as observed in the study of the interaction of actinomycin, derivatives and analogues with DNA at neutral pH (Müller and Crothers, 1968) in which the chromophore was shown to intercalate between base pairs. These studies have also established that actinomycin and its derivatives binds preferentially to G.C. containing sites and much more strongly to DNA than to RNA (Haselkorn, 1964; Müller and Crothers, 1968). The apparent number of binding sites was found larger for analogues without peptidic side chain (e.g. 2-aminophenoxazone-3) than for actinomycin. For this latter molecule several forms of the complex exist at equilibrium; this results from conformational change within the cyclic peptidic rings. Recent X-rays studies on a crystalline complex between deoxyguanosine and actinomycin D (Sobell *et al.*, 1971) can explain the specificity for G.C. containing sites. In these crystals the cyclic peptide residues have been found to interact with two deoxyguanosine molecules via specific hydrogen bonds: a strong hydrogen bond between the guanosine 2 amino group and the carbonyl oxygen of the L-threonine residue, and a weaker hydrogen bond between the guanosine N (3) ring nitrogen and the N–H group on this same L-threonine residue.

Aromatic amino acids and derivatives have been shown to interact with nucleic acids and polynucleotides via stacking between the aromatic ring and bases (or base-pairs) of the polynucleotide chain (Raszka and Mandel, 1971; Hélène *et al.*, 1971a, b). These molecules are charged at pH 7 and the contribution of electrostatic forces is indicated by the strong dependence of the extent of binding on the ionic strength (Hélène *et al.*, 1971a, b) and also on the sign of the electric charge carried by the small molecule: for analogues containing the same aromatic nucleus the effectiveness of binding decreases in the order amine > parent amino acid > aromatic acid derivative (Raszka and Mandel, 1971). P.m.r. and CD data show that they are conformational changes in the macromolecule which are induced by the binding of the small molecule (Raszka and Mandel, 1971; Hélène *et al.*, 1973). In the case of single-stranded neutral poly A the results can be unambiguously assigned to a destacking of the polynucleotide chains.

Extensive studies have been done concerning the interactions between salmon sperm DNA and various synthetic aromatic amines belonging to the following groups (Passero *et al.*, 1970):

(I) 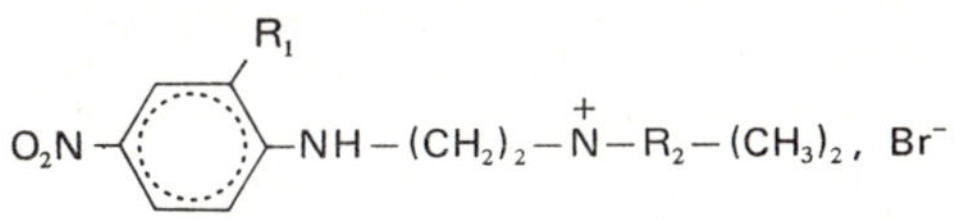

(II) 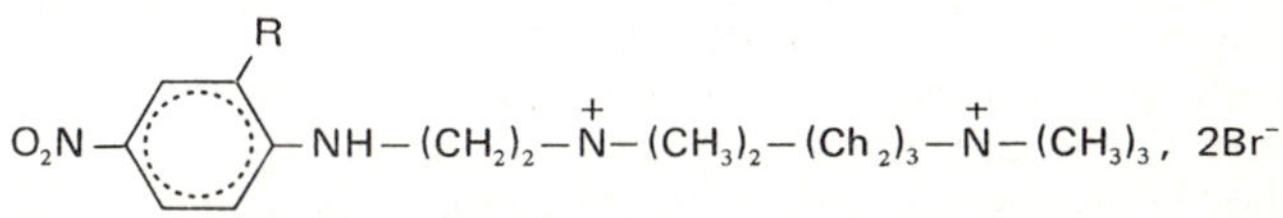

Besides the variation of the apparent binding constant on the ionic strength according to a simple competitive model, the contribution of the electrostatic forces is also indicated by the fact that the divalent type II molecules binds about ten times more strongly than the corresponding monovalent type I molecules. Intercalation of the aromatic ring between successive base pairs is strongly supported by p.m.r. data and by the observed increase in the intrinsic viscosity of the macromolecule on binding. When $R_2 = H$ and $R_1 = R = NO_2$ or CH_3 the relative intrinsic viscosity increment is smaller for type I molecules than for type II, suggesting that the monocation is not as extensively intercalated than the dication. Data obtained on molecules having a R_1 substituent of larger size

$$\left\{R_1 = CON\langle{R'_1 \atop R''_1}\right\}$$

show that there are steric requirements for the intercalation process to occur (Gabbay and De Paolis, 1971): for $R''_1 = H$ and $R'_1 = H$ or CH_3, intercalation occurs as shown by viscosity, p.m.r., absorbance and dichroism data. On the contrary, for $R'_1 = R''_1 = CH_3$ and $R'_1 = R''_1 = CH_2-CH_3$ the p.m.r. and viscosity data are rather in favour of external binding on the phosphodiester chain with the 4-nitroaniline ring freely tumbling in solution.

References

Akasaka, K., Sakoda, M. and Hiromi, K. (1970). *Biochem. biophys. Res. Comm.* **40**, 1239.

Applequist, J. and Damle, V. (1966). *J. Am. chem. Soc.* **88**, 3895.
Arav, R. and Wagner, K. G. (1970). *Eur. J. Biochem.* **13**, 267.
Armstrong, R. W., Kurucsev, T. and Strauss, U. P. (1970). *J. Am. chem. Soc.* **92**, 1260.
Arnott, S., Dover, S. D. and Wonacott, A. J. (1969). *Acta Cryst.* **B25**, 2192.
Arnott, S. (1970). *In* "Progress in Biophysics and Molecular Biology" (eds Butler, J. A. V. and Noble, D.). Pergamon, Oxford.
Atcher, E. and Felsenfeld, G. (1971). *Biopolymers* **10**, 1625.
Aylward, N. N. (1966). *J. Chem. Soc. sect. B.*, 627.
Baklagine, Y. G., Volkenstein, M. V. and Kondrasher, Y. D. (1963). *Zh. Struk. Khim.*, **7**, 399.
Bangerter, B. W. and Chan, S. I. (1968a). *Proc. natn. Acad. Sci. U.S.A.* **60**, 1144.
Bangerter, B. W. and Chan, S. I. (1968b). *Biopolymers* **6**, 983.
Bangerter, B. W. and Chan, S. I. (1969). *J. Am. chem. Soc.* **91**, 3910.
Barker, D. L. and Marsh, R. E. (1964). *Acta Cryst.* **17**, 1581.
Bauer, W. and Vinograd, J. (1970). *J. molec. Biol.* **54**, 281.
Bidet, R., Chambron, J. and Weill, G. (1971). *Biopolymers* **10**, 225.
Bittman, R. (1969). *J. molec. Biol.* **46**, 251.
Blake, A. and Peacocke, A. R. (1968). *Biopolymers* **6**, 1225.
Boyland, E. and Green, B. (1962). *Br. J. Cancer* **16**, 507.
Bradley, D. F. and Wolf, M. K. (1959). *Proc. natn. Acad. Sct. U.S.A.* **45**, 944.
Bradley, D. F. and Nash, H. A. (1968). *In* "Molecular Associations in Biology" (ed. Pullman, B.). Academic Press, New York and London.
Brahms, J., Michelson, A. M. and Van Holde, K. E. (1966). *J. molec. Biol.* **15**, 467.
Brahms, J. and Sadron, C. (1966). *Nature, Lond.* **212**, 1309.
Brahms, J., Maurizot, J. C. and Michelson, A. M. (1967). *J. molec. Biol.* **25**, 481.
Brahms, J. and Brahms, S. (1970). *In* "Biological Polyelectrolytes, vol. 4 (ed. Veis, A.), Marcel Dekker, New York.
Broom, A. P., Schweizer, M. P. and Ts'o, P. O. P. (1967). *J. Am. chem. Soc.* **89**, 3612.
Brown, R. A. (1966). *Archs. Biochem. Biophys.* **115**, 102.
Bugg, C. E. and Marsh, R. E. (1967). *J. molec. Biol.* **25**, 67.
Bugg, C. E., Thomas, J. M., Sundaralingam, M. and Rao, S. T. (1971). *Biopolymers* **10**, 575.
Bujard, (1968). *J. molec. Biol.* **33**, 503.
Caillet, J. (1968). *C.r.hebd. Séanc. Acad. Sci. Paris* **266 D**, 288.
Cantor, C. E. and Chin, W. W. (1968). *Biopolymers* **6**, 1745.
Chan, S. I., Schweizer, M. P., Ts'o, P. O. P. and Helmkam', G. K. (1964). *J. Am. chem. Soc.* **86**, 4182.
Chan, S. I. and Van Winkle, Q. (1969). *J. molec. Biol.* **40**, 491.
Chan, S. I. and Nelson, J. H. (1969). *J. Am. chem. Soc.* **91**, 168.
Claverie, P., Pullman, B. and Caillet, J. (1966). *J. theor. Biol.* **12**, 419.
Cohen, G. and Eisenberg, H. (1969). *Biopolymers* **8**, 45.
Craig, A. M. and Isenberg, I. (1970a). *Biopolymers* **9**, 689.
Craig, A. M. and Isenberg, I. (1970b). *Proc. natn. Acad. Sci. U.S.A.* **67**, 1337.
Crawford, L. V. and Waring, M. J. (1967a). *J. molec. Biol.* **25**, 23.
Crawford, L. V. and Waring, M. J. (1967b). *J. gen. Virol.* **1**, 387.
Danner, J. and Helmkamp, G. K. (1971). *Biochim. biophys. Acta* **232**, 227.
Davies, R. R. (1967). *A. Rev. Biochem.* **36**, 321.
De Voe, H. and Tinoco, I. jr. (1962). *J. molec. Biol.* **4**, 500.

Dimicoli, J. L. and Hélène, C. (1971). *Biochimie* **53**, 331.
Dimicoli, J. L. and Hélène, C. (1972) *J. Am. Chem. Soc.* **95**, 1036
Domiani, A., De Santis, P., Giglo, E., Liquori, A. M., Puliti, R. and Ripamonti, A. (1965). *Acta. Cryst.* **19**, 340.
Dourlent, M., Thrierr, J. C., Brun, F. and Leng, M. (1970). *Biochem. biophys. Res. Comm.* **41**, 1590.
Dourlent, M. and Hélène, C. (1971). *Eur. J. Biochem.* **23**, 86.
Drummond, D. S., Simpson-Gildemeister, V. F. W. and Peacocke, A. R. (1965). *Biopolymers* **3**, 135.
Eisenberg, H. and Felsenfeld, G. (1967). *J. molec. Biol.* **30**, 17.
Ellerton, N. F. and Isenberg, I. (1969). *Biopolymers* **8**, 767.
Farquhar, E. L., Downing, M. and Gill, S. J. (1968). *Biochemistry* **7**, 1224.
Felsenfeld, G., Davies, D. R. and Rich, A. (1957), *J. Am. chem. Soc.* **79**, 2023.
Felsenfeld, G. and Miles, T. H. (1967). *A. Rev. Biochem.* **36**, 407.
Flory, P. J. (1953). *In* "Principles of Polymer Chemistry". Cornell University Press, Ithaca.
Flory, P. J. (1969). *In* "Statistical Mechanism of Chains Molecules". Interscience Publishers, New York.
Flory, P. J. and Jernigan, R. L. (1965). *J. Chem. Phys.* **42**, 3509.
Fuller, W. and Waring, M. J. (1964). *Ber. Bunsenges. Physik. Chem.* **68**, 805.
Gabbay, E. J. and De Paolis, A. (1971). *J. Am. chem. Soc.* **93**, 562.
Gellert, M., Lipsett, M. N. and Davies, D. R. (1962). *Proc. natn. Acad. Sci. U.S.A.* **48**, 2013.
Gill, S. J., Martin, D. B. and Downing, M. (1963). *J. Am. chem. Soc.* **85**, 706.
Gottschalk, E., Kopp, E. and Lezuis, A. G. (1971). *Eur. J. Biochem.* **24**, 168.
Green, B. and McCarter, J. A. (1967). *J. molec. Biol.* **29**, 447.
Green, G. and Malher, H. R. (1970). *Biochemistry* **9**, 368.
Gulik, A., Inone, H. and Luzzati, V. (1970). *J. molec. Biol.* **53**, 221.
Hamlin, R. M., Lord, R. C. and Rich, A. (1965). *Science, N.Y.* **148**, 1734.
Hammes, G. G. and Hubbard, C. D. (1966). *J. chem. Phys.* **70**, 2889.
Hammes, G. G. and Park, A. C. (1968). *J. Am. chem. Soc.* **90**, 4151.
Haschemeyer, A. E. V. and Rich, A. (1967). *J. molec. Biol.* **27**, 369.
Haschemeyer, A. E. V. and Sobell, H. M. (1965). *Acta Cryst.* **18**, 525.
Haschemeyer, A. E. V. and Sobell, H. M. (1965). *Acta CVyst.* **19**, 125.
Haselkorn, R. (1964). *Science, N.Y.* **143**, 682.
Hélène, C., Dimicoli, J. L., Borazan, H. N., Durand, M., Maurizot, J. C. and Toulmé, J. J. (1973). *In* "Conformation of Biological Molecules and Polymers", p. 361. Jerusalem.
Hélène, C., Dimicoli, J. L. and Brun, F. (1971a). *Biochemistry* **10**, 3802.
Hélène, C., Montenary-Garestier, Th. and Dimicoli, J. L. (1971b). *Biochim. biophys. Acta* **254**, 349.
Helmkamp, G. K. and Kondo, N. S. (1967). *Biochim. biophys Acta* **145**, 27.
Helmkamp, G. K. and Kondo, N. S. (1968). *Biochim. biophys. Acta* **157**, 242.
Holcombe, D. N. and Tinoco, I. Jr. (1965). *Biopolymers* **3**, 121.
Hoogsteen, K. (1959). *Acta Cryst.* **12**, 822.
Hoogsteen, K. (1963). *Acta Cryst.* **16**, 907.
Hoogsteen, K. (1968). *In* "Molecular Association in Biology" (ed. Pullman, B.), p. 21. Academic Press, New York and London.
Hospital, M. and Busetta, B. (1969). *C.r. hebd. Séanc. Acad. Sci. Paris* **268**, 1232.

Howard, F. B., Frazier, J., Lipsett, M. N. and Miles, H. T. (1964). *Biochem. biophys. Res. Comm.* **17**, 93.
Howard, F. B., Frazier, J. and Miles, T. H. (1971). *J. biol. Chem.* **246**, 7073.
Howard, F. B., Frazier, J., Singer, M. F. and Miles, H. T. (1966). *J. molec. Biol.* **16**, 415.
Hruska, F. E., Bell, C. L., Victor, T. A. and Danyluk, S. S. (1968). *Biochemistry* **7**, 3721.
Huang, W. M. and Ts'o, P. O. P. (1966). *J. molec. Biol.* **16**, 523.
Hui, Bon Hoa, G., Douzou, P. and Michelson, A. M. (1966). *Biochim. biophys. Acta* **182**, 334.
Inners, D. L. and Felsenfeld, G. (1970) *J. molec. Biol.* **50**, 373.
Isenberg, I., Baird, S. L. and Bersohn, R. (1967). *Biopolymers* **5**, 477.
Isenberg, I., Baird, S. L. and Bersohn, R. (1969). *Ann. N.Y. Acad. Sci.* **153**, 789.
Jardetsky, O. (1965). *Biopolymer Symp.* **1**, 501.
Jeffrey, G. A. and Kinoshita, Y. (1963). *Acta Cryst.* **16**, 20,
Job, P. (1928). *Ann. Chimie.* **9**, 113.
Kasarda, D. D. (1970). *Biochim. biophys. Acta* **217**, 535.
Katz, L. and Penman, S. (1966). *J. molec. Biol.* **15**, 220.
Katz, L., Tomita, K. and Rich, A. (1966). *Acta Cryst.* **21**, 754.
Kearns, D. R., Patel, D., Shulman, R. G. and Yamane, T. (1971). *J. molec. Biol.* **61**, 265.
Kreishman, G. P., Chan, S. I. and Bauer, W. (1971). *J. molec. Biol.* **61**, 45.
Kuchler, E. and Derkosch, J. (1966). *Z. Naturf.* **216**, 209.
Kyogoku, Y., Lord, R. C. and Rich, A. (1966). *Science, N.Y.* **154**, 518.
Kyogoku, Y., Lord, R. C. and Rich, A. (1967a). *J. Am. chem. Soc.* **89**, 496.
Kyogoku, Y., Lord, R. C. and Rich, A. (1967b). *Proc. natn. Acad. Sci. U.S.A.* **57**, 250.
Leng, M. and Felsenfeld, G. (1966). *J. molec. Biol.* **15**, 455.
Lerman, L. S. (1961). *J. molec. Biol.* **3**, 18.
Lesko, A., Smith, A., Ts'o, P. O. P. and Umans, R. S. (1968). *Biochemistry* **7**, 434.
Lewin, S. (1964). *J. chem. Soc.* **5**, 792.
Li, H. J. and Crothers, D. M. (1969). *J. molec. Biol.* **39**, 461.
Lippset, M. N. (1960). *Proc. natn. Acad. Sci. U.S.A.* **46**, 445.
Liquori, A. M., Delerma, B., Ascoli, F., Botre, C. and Trasciatti, M. (1962). *J. molec. Biol.* **5**, 521.
Löber, G. and Achtert, G. (1969). *Biopolymers* **8**, 596.
Lord, R. C. and Thomas, G. J. (1967). *Biochim. biophys. Acta* **142**, 1.
Lunell, S. and Sperber, G. (1967). *J. chem. Phys.* **46**, 2119.
Margar, M. E. and Steiner, R. F. (1970). *Biochim. biophys. Acta* **224**, 80.
Magar, M. E., Steiner, R. F. and Kolinski, R. (1971). *Biophys. J.* **11**, 387.
Mantsch, H. M. and Smith, I. C. P. (1972). *Biochem. biophys. Res. Comm.* **46**, 808.
Massoulie, J. (1968). *Eur. J. Biochem.* **3**, 428.
Mathews, F. S., and Rich, A. (1964). *J. molec. Biol.* **8**, 89.
Maurizot, J. C., Blicharski, J., and Brahms, J. (1971). *Biopolymers* **10**, 1429.
Michelson, A. M., Massoulie, J. and Gushlbauer, W. (1967). *Prog. Nucleic. Acid. Res. molec. Biol.* **6**, 83.
Michelson, A. M. and Monny, C. (1966). *Proc. natn. Acad. Sci. U.S.A.* **56**, 1528.
Miles, H. T. (1964). *Proc. natn. Acad. Sci. U.S.A.* **51**, 1104.
Millar, D. B. (1968). *Biochim. biophys. Acta.* **166**, 628.
Millar, D. B. and McKenzie, M. (1970). *Biochim. biophys. Acta* **204**, 82.

Millar, D. B. and Steiner, R. F. (1965). *Biochim. biophys. Acta* **102**, 571.
Millar, J. H. and Sobell, H. M. (1967). *J. molec. Biol.* **24**, 345.
Montenay-Garestier, T. and Hélène, C. (1971). *Biochemistry* **10**, 300.
Müller, W. and Crothers, D. M. (1968). *J. molec. Biol.* **35**, 251.
Nagata, C., Kodama, M., Tayashira, Y. and Imamura, A. (1966). *Biopolymers* **4**, 409.
Nakano, N. I. and Igarashi, S. J. (1970). *Biochemistry* **9**, 577.
Nash, H. A. and Bradley, D. F. (1966). *J. chem. Phys.* **45**, 1380.
Neville, D. M. and Davies, D. R. (1966). *J. molec. Biol.* **17**, 57.
Newark, R. A. and Cantor, C. R. (1968). *J. Am. chem. Soc.* **90**, 5010.
O'Brien, E. J. (1963). *J. molec. Biol.* **7**, 107.
O'Brien, E. J. (1966). *J. molec. Biol.* **22**, 377.
Paoletti, J. and Le Pecq, J. B. (1971a). *J. molec. Biol.* **55**, 43.
Paoletti, J. and Le Pecq, J. B. (1971b). *Biochimie* **53**, 969.
Passero, F., Gabbay, E. J., Gaffney, B. and Kurucsev, T. (1970). *Macromolecules* **3**, 158.
Pitha, J., Jones, N. and Phithova, P. (1966). *Can. J. Chem.*, **44**, 1045.
Pitha, P. M., Huang, W. M. and Ts'o, P. O. P. (1968). *Proc. natn. Acad. Sci. U.S.A.* **61**, 332.
Pochon, F. and Michelson, A. M. (1965). *Proc. natn. Acad. Sci. U.S.A.* **53**, 1425.
Poland, D., Vournakis, J. N. and Scheraga, H. A. (1966). *Biopolymers* **4**, 223.
Pörschke, D. (1971). *Biopolymers* **10**, 1989.
Pörschke, D. and Eigen, M. (1971). *J. molec. Biol.* **62**, 361.
Pritchard, N. J., Blake, A. and Peacocke, A. R. (1966). *Nature, Lond.* **212**, 1260.
Pullman, B., Claverie, P. and Caillet, J. (1966a). *Proc. natn. Acad. Sci. U.S.A.* **55**, 904.
Pullman, B., Claverie, P. and Caillet, J. (1966b). *J. molec. Biol.* **22**, 373.
Raszka, M. and Mandel, M. (1971). *Proc. natn. Acad. Sci. U.S.A.* **68**, 1190.
Raszka, M. and Kaplan, N. (1972). *Proc. natn. Acad. Sci. U.S.A.* **69**, 2025.
Ravindranathan, R. V. and Miles, H. T. (1965). *Biochim. biophys. Acta* **94**, 603.
Rein, R. and Harris, R. A. (1966), *J. chem. Phys.* **45**, 1797.
Rich, A. (1957). *In* "The Chemical Basis of Heredity" (eds MacElroy, W. D. and Glass, B.). Johns Hopkins Press, Blatimore.
Richards, E. G., Flessel, C. P. and Fresco, J. R. (1963). *Biopolymers* **1**, 431.
Rifkind, J. M. and Eichorn, G. L. (1970). *Biochemistry* **9**, 1753.
Rogers, G. T., Ulbricht, T. L. V. and Szer, W. (1967). *Biochem. biophys. Res. Comm.* **27**, 372.
Roth, D. and Manjon, M. L. (1969). *Biopolymers* **7**, 695.
Sakoda, M., Hiromi, K. and Akasaka, K. (1971). *Biopolymers* **10**, 1003.
Sander, C. and Ts'o, P. O. P. (1970). *Biopolymers* **9**, 765.
Sarkar, P. K. and Yang, J. T. (1965). *Biochem. biophys. Res. Comm.* **20**, 346.
Sasisekharan, V., Lakshminarayanam, A. V. and Ramachandran, G. N. (1967). *In* "Conformations of Biopolymers" (ed. Ramachandran, G. N.), vol. 2. Academic Press, London and New York.
Sasisekharan, V. and Lakshminarayanan, A. V. (1969). *Bioplymers* **8**, 475.
Sasisekharan, V. and Sigler, P. B. (1965). *J. molec. Biol.* **12**, 296.
Saucier, J. M., Festy, B. and Le Pecq, J. B. (1971). *Biochimie* **53**, 973.
Scheit, K. H. and Faerber, P. (1971). *Eur. J. Biochem.* **24**, 385.
Schmechel, D. E. V. and Crothers, D. M. (1971). *Biopolymers* **10**, 465.
Schmitz, K. S. and Schurr, J. M. (1971). *Biopolymers* **10**, 2075.

Schwarz, G. (1970). *Eur. J. Biochem.* **12**, 442.
Schwarz, G. and Balthasar, W. (1970). *Eur. J. Biochem.* **12**, 461.
Schwarz, G., Close, S. and Balthasar, W. (1970). *Eur. J. Biochem.* **12**, 454.
Schweizer, M. P., Broom, A. P., Ts'o, P. O. P. and Hollis, D. P. (1968). *J. Am. chem. Soc.* **90**, 1042.
Scruggs, R. L. and Ross, P. D. (1970). *J. molec. Biol.* **47**, 29.
Shoup, R. R., Miles, H. T. and Becker, E. D. (1966). *Biochem. biophys. Res. Comm.* **23**, 194.
Singh, H. and Hillier, B. (1971). *Biopolymers* **10**, 2445.
Sobell, H. M., Jain, S. C., Sakore, T. D. and Nordman, C. E. (1971). *Nature, New Biology* **231**, 200.
Sobell, H. M., Tomita, K. and Rich, A. (1963). *Proc. natn. Acad. Sci. U.S.A.* **49**, 885.
Solie, T. N. and Schellman, J. A. (1968). *J. molec. Biol.* **33**, 61.
Steiner, R. F. (1970). *Biochemistry* **9**, 4268.
Steiner, R. F. (1971). *Biochim. Biophys. Acta* **247**, 22.
Steiner, R. and Millar, D. B. (1970). *In* "Biological Polyelectrolytes", vol 3 (ed. Veis, A.). Marcel Dekker, New York.
Stevens, C. and Felsenfeld, G. (1964). *Biopolymers* **2**, 293.
Stone, A. L. (1967), *Biochim. biophys. Acta* **148**, 193.
Stone, A. L. and Bradley, D. F. (1961). *J. Am. chem. Soc.* **83**, 3627.
Stone, A. L. and Bradley, D. F. (1967). *Biochim. biophys. Acta.* **148**, 173.
Stone, A. L., Childers, L. G. and Bradley, D. F. (1970). *Trans. Farad. Soc.* **66**, 3081.
Strauss, U. P. and Ander, P. (1962). *J. Phys. Chem.* **66**, 2635.
Sundaralingam, M. (1966). *Acta Cryst.* **21**, 495.
Sundaralingam, M. (1969). *Biopolymers* **7**, 821.
Sundaralingam, M., and Jensen, L. H. (1965). *J. molec. Biol.* **13**, 914.
Szer, W. (1970). *Ann. N.Y. Acad. Sci.* **171**, 801.
Szer, W. and Sugar, D. (1963). *Acta biochim. Polon.* (English trans.) **10**, 219.
Thiele, D. and Guschlbauer, W. (1971). *Biopolymers* **10**, 143.
Thomes, J. C., Weill, G. and Daune, M. (1969). *Biopolymers* **8**, 647.
Thrierr, J. C., Dourlent, M. and Leng, M. (1971). *J. molec. Biol.* **58**, 815.
Thrierr, J. C. and Leng, M. (1969). *Biochem. biophys. Acta* **182**, 575.
Thrierr, J. C. and Leng, M. (1972). *Biochim. biophys. Acta* **272,** 238.
Tomita, T., Katz, L., and Rich, A. (1967). *J. molec. Biol.* **30**, 545.
Tritton, T. R. and Mohr, S. C. (1971). *Biochem. biophys. Res. Comm.* **45**, 1240.
Ts'o, P. O. P. (1969). *Ann. N.Y. Acad. Sci.* **158**, 256.
Ts'o, P. O. P. (1970). *In* "Fine Structure of Proteins and Nucleic Acids" (eds Fasmann, G. and Timasheff, N.). Marcel Dekker, New York.
Ts'o, P. O. P. and Chan, S. I. (1964). *J. Am. chem. Soc.* **86**, 4176.
Ts'o, P. O. P. and Huang, W. M. (1968). *Biochemistry* **7**, 2954.
Ts'o, P. O. P., Kondo, N. S., Robins, R. K. and Broom, A. D. (1969). *J. Am. chem. Soc.* **91**, 5625.
Ts'o, P. O. P., Melvin, I. S. and Olson, A. C. (1963). *J. Am. chem. Soc.* **87**, 5241.
Ts'o, P. O. P., Rapaport, S. A. and Bollum, F. J. (1966). *Biochemistry* **5**, 4153.
Ts'o, P. O. P. and Schweizer, M. P. (1968). *Biochemistry* **7**, 2963.
Tubbs, R. K., Ditmar, W. E. and Van Winkle, Q. (1964). *J. molec. Biol.* **9**, 545.
Van Holde, K. E. and Rosseti, G. P. (1967). *Biochemistry* **6**, 2189.
Vinograd, J., Lebowitz, J. and Watson, R. (1968). *J. molec. Biol.* **33**, 173.

Wagner, K. G. (1969). *Euro. J. Biochem.* **10**, 261.
Wagner, K. G. and Arav, R. (1968). *Biochemistry* **7**, 1771.
Waring, M. (1970). *J. molec. Biol.* **54**, 247.
Warner, R. C. (1957). *J. biol. Chem.* **229**, 711.
Warshaw, M. M. and Tinoco, I. Jr. (1966). *J. molec. biol.* **19**, 29.
Weill, G. and Calvin, M. (1963). *Biopolymers* **1**, 401.
Witz, J. and Luzzati, V. (1965). *J. molec. Biol.* **11**, 620.
Woese, C. (1968). *Proc. natn. Acad. Sci. U.S.A.* **59**, 110.
Wulf, K., Wolf, H. and Wagner, K. G. (1970). *Biochem. biophys. Res. Comm.* **39**, 870.
Zmudzka, B. and Shugar, D. (1971). *Acta biochim. Pol.* **18**, 321.

CHAPTER 19

The Polymer and Salt-induced Condensation of DNA†

L. S. LERMAN‡

Department of Molecular Biology, Vanderbilt University, Nashville, Tennessee

I. Polymer and Salt-induced Condensation

The presence of simple uncharged polymers in low concentration has little, if any, effect on the properties of DNA in solution. However, when the polymer concentration exceeds a characteristic critical limit, the solutions change substantially (Lerman, 1971a; Lerman, 1971b; Jordan *et al.*, 1972), and if the DNA concentration is sufficiently high, a visible turbidity or precipitate may form. Since the polymer concentration necessary for these changes is dependent on salt concentration, solutions of uniform polymer concentration may be either above or below the critical condition, depending on the salt. For low concentrations of high molecular weight DNA($a_{260,\ 1\ \mathrm{cm}} \cong 0{\cdot}2$) the change in salt or polymer concentration needed to cross the transition zone is exceedingly small, as is characteristic of a phase change or a highly cooperative process. The effect is fully reversible, with no indication of denaturation either in the presence of above-critical polymer and salt or after returning to

† This work was supported by grants from the National Science Foundation (GB 30734) and from the National Institute of Health (GM 13767).

‡ Fellow of the John Simon Guggenheim Memorial Foundation 1971–72.

a normal solution. It will be convenient to refer to the polymer-and-salt-induced transition by the acronym, ψ.

The phenomenon has seemed interesting because DNA changes from its extended, solution configuration into a compact state that is conspicuously non-fibrous. Even when DNA is brought into over-critical ψ conditions under laminar flow with a reasonably high rate of shear, fibres are generally not formed—a marked contrast to the propensity toward fibre formation in alcohol-containing solutions. The critical dependence on salt is reminiscent of the (less) sharp decline in solubility of chromatin as the salt concentration approaches isotonicity. The absence of any basis for attractive interactions between the polymers and DNA suggests that the forces by which the compact state is imposed are gentle and non-specific, and therefore that the changes, though unfamiliar, are readily accessible to the DNA molecule. The sharp response to small environmental changes suggests also a property appropriate to a biological control mechanism.

Most of our studies have been carried out with poly(ethylene oxide) of nominal molecular weight 6000–7500, corresponding to a degree of polymerisation of 136–170, but similar effects are given by other sizes of this polymer (at different concentrations) and by poly(vinyl pyrrolidinone), an uncharged polymer that is chemically quite different. A negative polyanion, poly(acrylate), behaves similarly, but its examination has been less extensive. We find no substantial differences in the properties of the system if the commercial poly(ethylene oxide) (Union Carbide Corporation, Carbowax and Polyox) is charcoal treated to remove u.v.-absorbing impurities, or purified by precipitation, or examined in the presence of EDTA. In the following, poly(ethylene oxide) will be abbreviated, PEO.

The water solubility of PEO is indefinitely high; clear solutions containing in excess of 300 mg/ml of the 6000–7500 MW material are easily prepared. The dependence of its intrinsic viscosity on molecular weight in water indicates that polymer–water contacts are favoured over polymer–polymer contacts. As a macromolecule, PEO in the relevant concentrations diminishes the activity of water only very slightly. Salt concentrations in the range of the transition effect a slight contraction of the randomcoil, and θ conditions are obtained only at higher concentrations, 0·45 M K_2SO_4, 35° (Bailey and Callard, 1959). PEO exhibits an exceedingly weak affinity for polycyclic aromatic molecules in aqueous solution (Okubo and Ise, 1969; Green and Lerman, unpublished), $-\Delta F^0 \cong 0{\cdot}6$ kcal/mol if 10 monomer residues constitute a binding site for ethidium.

II. Volume Exclusion and Phase Separation

Thermodynamic principles pertaining to the interactions of two macromolecules present in the same solution relate the separation of the solution into two phases to weak, short-range repulsions. The large dimensions of randomcoil polymers limit solutions to a relatively small number of solute molecules, and require that most of the space containing solvent lie within the domain of one polymer molecule or another; also, intermolecular contacts will be frequent. For small numbers of molecules the free energy of mixing is easily overwhelmed by the effect of a very small positive free energy of contact between the polymers. In general, homogeneous solutions cannot be made containing appreciable concentrations of two polymers in a solvent in which each may be separately quite soluble, because of these interactions (Flory, 1953). Thus, a mixture containing DNA and another water-soluble polymer can be expected to separate spontaneously into two distinct phases, one rich in DNA and the other rich in the second polymer. Two-phase aqueous systems of water-soluble polymers have become familiar through their application by Albertsson (1971) for macromolecular separations. It may be noted that the phase separation does not depend on dehydration, that is, competition for binding of solvent molecules or lowering of solvent activity by the polymer, but rather it depends on competition for the space of the solution and the lack of attractive interactions when polymer molecules collide. It may further be supposed that escape from the unfavourable free energy of interaction might be provided by a change in conformation, if the free energy of the conformational change is also small and a sufficient diminution of unfavourable contacts results. While it is not yet clear that these principles completely account for the changes in DNA in polymer solution, as discussed in the following sections, they provided the basis for initiating these studies and remain compatible with our observations. No preferable interpretation has yet been devised. Since, in particular, the critical role of salt in the transition is not included, the three component theory may be a useful approximation when the salt concentration is high.

In a solution containing DNA and a lower molecular weight polymer there is an asymmetry in the spatial requirements of the two macromolecules. Because of its stiffness a single DNA molecule may block out an enormous domain in the solvent. For example, the root-mean-square radius (the radius of gyration) of a T4 DNA molecule defines a sphere of about 2600 nm in diameter of which only 10^{-5} of the mass is DNA (the remainder, solvent), and yet 70% of the DNA molecule can be expected to extend into the space outside this average sphere. Correspondingly, the average distances between nearby (though obviously not necessarily sequentially contiguous) segments of the DNA helix near the centre of the sphere, its densest region, will be large, roughly 45 nm. We may compare these dimensions with those of PEO of nominal molecular weight near 6800, the polymer used in most of the following experiments. Because of both the much smaller molecular weight and the greater flexibility of PEO, its root-mean-square radius is in the vicinity of 3·2 nm, defining a sphere which contains only 90% water. Thus, molecules of PEO can be expected to move almost freely within the domain of a DNA molecule, while the DNA helix finds each PEO a virtually impenetrable blob because of the width of the double helix. The system can be seen to resemble the mixture of serum albumin and hyaluronic acid, which has been studied experimentally by Ogston and his collaborators (see Laurent and Ogston, 1963). Ogston (1962) has extended the theory of ternary solutions to consider the dependence of phase separation on the interaction between the two such polymers based on their competition for solvent space and their non-attractive contacts, which define an excluded volume. For the present purposes, the excluded volume can be calculated on the basis of a hard rod and sphere, since the radius of curvature of the rod (DNA) and the interstitial spaces are large compared to the diameter of the sphere (PEO). As PEO enters the space of a DNA molecule, it is assumed to move freely until one of its peripheral segments encounters the double helix. The distance of that encounter would represent the closest approach between the two molecules and will be seen to define a cylindrical volume, including and surrounding the double helix, into which the centre of the PEO molecule cannot enter. The total volume will depend on the length of the DNA and the sum of the radius of the double helix and a suitably representative radius of the PEO randomcoil. Obviously, the volume excluded for a polymer of molecular weight 6800 will be a small fraction of the total volume of the solvent, but it can be supposed that at a sufficiently high concentration of the polymer, the thermodynamic consequences of exclusion may not be negligible.

The condition for the stability of a ternary solution is given by the relation $\mu_{22}\mu_{33} - \mu_{23}^2 \geqq 0$, where μ_{22} is the partial derivative of the chemical potential of PEO with respect to its own molal concentration, μ_{33} is the partial derivative of the chemical potential of DNA with respect to its molal concentration and μ_{23} is the partial derivative of the chemical potential of either of the two macromolecules with respect to the molal concentration of the other. Following Ogston, and ignoring second virial coefficients here, we may take $\mu_{22} \cong \mathrm{RT}/m_2$ and $\mu_{33} \cong \mathrm{RT}/m_3$, where m_2 and m_3 are the molal concentrations of PEO and DNA respectively. The value of μ_{23} is estimated in terms of a partition coefficient representing the distribution of one polymer between two phases in equilibrium—one phase being the mixed solution of both polymers, and another otherwise identical phase in which the second polymer is absent. Inside a chamber defined by a membrane permeable to PEO and containing DNA, the equilibrium PEO concentration, as indicated by the partition coefficient, *K*, would be slightly lower than in the outside, DNA-free solution, because of the volume inside from which PEO is excluded. Then $\mu_{23} = (\mathrm{RT} \ln K)/m_3$.

When the DNA molecules are very long, we can recognize two somewhat different ways of treating the excluded volume. In both we estimate K, and consequently μ_{23}, from geometrical considerations alone, and ignore any details of the intermolecular interaction except the effect of very short range repulsion.

First, in the more obvious sense, the configuration of individual DNA molecules need not be considered; the solution is taken as homogeneous. If the volume (cm^3) of the cylinder of exclusion as determined by the total length of dissolved DNA (end effects are negligible) per millilitre of solution is ϵ, its value will be proportional to the nucleotide concentration, c. Where r_h is the radius of the double helix, r_p is the effective radius of the polymer randomcoil (dimensions in nm), N is Avogadro's number, c is in nucleotide equivalents per litre, and the helix increases 0·17 nm in length per nucleotide: $\epsilon = 0{\cdot}17\pi cN(r_h+r_p)^2 \times 10^{-24}$. Then $K = 1+\epsilon$ and since ϵ is small, $\ln K \simeq \epsilon$. The concentration of polymer which is critical for phase separation is then defined by the stability relation such that

$$\frac{1}{m_2}\cdot\frac{1}{m_3} - \left(\frac{\ln K}{m_3}\right)^2 = 0 \quad \text{or } m_2 = m_3/\epsilon^2.$$

Where X is the number of nucleotides per DNA molecule and b includes the constants in the expression for ϵ, $m_2 = b/cX\,(r_h+r_p)^4$.

The critical polymer concentration for phase separation from a very dilute solution (260 nm absorbance, 1 cm, of about 0·07) of moderately high molecular weight DNA can be calculated, using a value of r_p inferred from the intrinsic viscosity function. If we introduce the following empirical numerical values, $r_h = 1{\cdot}2$ nm, $r_p = 3{\cdot}2$ nm, $c = 10^{-5}$ N, $X = 10^5$, and PEO molecular weight of 6800, m_2, converted to weight/volume solution is found to be 145 mg/ml. Although the experimental value is only in order-of-magnitude agreement, it should be noted that the theory is drastically simplified, the approximations are crude, and no arbitrary parameters have been introduced. An additional complication may be inferred; the effect of the second virial coefficient of DNA, which has been neglected, would be to raise the critical polymer concentration by an amount independent of the DNA concentration.

Ogston's relations, again simplified for the present calculation, also permit estimation of the DNA concentration in the denser phase that would be in equilibrium with 145 mg/ml PEO, as above. The result is about $1{\cdot}8 \times 10^{-5}$ M, or 610 mg/ml DNA, together with about $1{\cdot}4 \times 10^{-7}$ M, or 10^{-3} mg/ml PEO in the DNA phase. We find about 230 mg/ml DNA in equilibrium with 150 mg/ml PEO (Maniatis, 1971), experimentally. Again the precise values are unimportant; the significance of the estimate is that an exceedingly high DNA concentration is expected in the dense phase—a concentration that is incompatible with the worm-like randomcoil configuration of double helical DNA in ordinary dilute solutions. Although concentrations like this can be realised in fibres, the dense phases in these experiments are not fibrous, as has been noted above.

III. Monomolecular Condensation

In a dilute solution of large DNA molecules the spatial distribution of helical segments is obviously quite inhomogeneous—there is a high, essentially macroscopic concentration near the centre of each randomcoil, and essentially zero elsewhere. We may speculatively regard a single DNA molecule as large enough to constitute a macroscopic phase, and consider the consequence of a critical polymer concentration near the centre. One possible meaning of phase separation with respect to a single DNA molecule is clarified by regarding the molecule as an assembly of a large number of relatively independent helical segments that are lightly tethered. With these segments as the elements of the DNA phase, phase separation would correspond to their clustering within a smaller volume of the solvent from which PEO would be excluded. Clustering of the segments would be formally equivalent to a conformational change of the DNA molecule as a whole. We can ask whether the conditions for phase separation can be met in terms of the average DNA concentration within the central region of the molecule, as defined by the root-mean-square radius of the worm-like chain. If the volume within the RMS sphere is V_{RMS}, and the excluded volume is ϵ, the critical molal concentration for marginal stability of the mixture is $\mathrm{m}_2 = \mathrm{m}_3 V^2_{\mathrm{RMS}}/\epsilon^2$. The RMS volume can be calculated from the dimensions of the statistically equivalent randomcoil corresponding to the worm-like chain model of DNA, for which the RMS radius, R, is given to a satisfactory approximation by $R=(Lp/3)^{1/2}$ where p is the persistence length, a parameter reflecting the stiffness of DNA, and L is the contour length of the entire DNA molecule. The excluded volume can be taken as the volume of a cylinder with the contour length of that part of the DNA lying within R, 30·5% of $0{\cdot}34x/2$, and a cross-sectional radius given, as before, by the sum of the radius of the DNA helix, r_h, and the root-mean-square radius of PEO, r_p. The remaining parameter is that of the molal concentration of DNA, which can be taken to be 1 DNA molecule per V_{RMS}. The weight concentration (mg/g solvent) of PEO necessary to reach the critical stability limit is then $W_2 = 11p^{3/2}\,\mathrm{M}_2/(r_h+r_p)^4 \times V_2$. When plausible values of the parameters are introduced—45 nm for the persistence length (Triebel *et al.*, 1971), 6000 for the molecular weight of PEO, and 180 000 as a compromise value for the number of nculeotide pairs in T4 DNA, the effective monomeric DNA concentration near the centre will be about 25×10^{-6} nucleotide equivalents/l (N), corresponding to a 260 nm absorbance for 1 cm of a uniform solution of the same density of about 0·17. Since the result will be extremely sensitive to the choice of the value for R_e, the radius at which PEO behaves as a hard sphere in contact with the DNA helix, we can take as the upper limit the maximum plausible value, 1/2 the largest diameter of the randomcoil (at this distance the PEO could rotate freely near the helix) or, as the lower limit, the RMS radius of PEO. These are 5·5 and 3·0 nm, respectively. The critical values of w_2, corrected to weight per unit volume are then found to be 126 and 19·5 mg/ml respectively.

Experiments corresponding to unimolecular changes are discussed in section VII; the PEO concentrations are within the expected range.

IV. Interaction Properties

While substantial experimental data for comparison with these functions are not yet available, measurements have been made under particular conditions and in preliminary experiments on the dependence of the critical PEO concentration on its molecular weight. In 1 M sodium the critical polymer concentration, estimated from the low speed sedimentability of commercial calf thymus DNA ($a_{260,\ 1\ \mathrm{cm}} \cong 0{\cdot}15$), varies from 135 mg/ml to 6 mg/ml as the molecular weight of PEO varies from 600 to 2×10^5, and conforms reasonably well to an inverse proportionality to the square root of the molecular weight (the nominal values specified by the manufacturer). Despite the heterogeneity of the unfractionated, eucaryotic DNA, the critical concentration is sharply defined. Higher polymer concentrations are required if the sodium concentration is lower. Although both the equations for bulk and unimolecular phase separation might seem to suggest an asymptotic inverse first-power dependence on molecular weight on the basis of the properties of infinitely dilute solutions of a monodisperse polymer under θ conditions, the system obviously does not fit those restrictions. Literature data on PEO are also somewhat difficult; both a 0·3 and a 0·8 power dependence of the intrinsic viscosity on molecular weight (for low MW) in water have been reported.

Observations on the unimolecular collapse of T4 DNA in two samples of poly(vinyl pyrrolidinone) (see section VII) are also compatible with an inverse square root dependence on the polymer molecular weight.

With a finite DNA concentration, but not too large, the formation of a dense phase may proceed first by the second process, unimolecular collapse followed by coalescence of the collapsed, or partially collapsed molecules into bigger units.

The diminution of the volume from which PEO might be excluded if the DNA were to collapse to a compact form can be seen in the corresponding calculation for the T4 DNA molecule folded into a sphere with about the same volume as the head of T4 phage. The helices are then in sufficiently close proximity that PEO cannot penetrate the array (compare the minute PEO concentration expected in the dense DNA phase, section II). The excluded volume than includes the total volume of the sphere plus a shell around it with the thickness of the RMS radius of PEO. Where the previous excluded volume, following our previous convention of considering only 30·5% of the molecule, was $0{\cdot}93 \times 10^6$ nm^3 in the compact sphere, the total excluded volume, which depends on the entire molecule, is $1{\cdot}3 \times 10^4$ nm^3, only 1·4% as large.

V. The Effect on Water Activity

Since the critical polymer concentration seems large, regarded as a weight or volume fraction of the solution, it is of interest to consider the thermodynamic effect of the presence of the polymer on water, particularly, whether any dehydration of DNA may be expected. With respect to the extreme (and unsuitable) supposition that the polymer behaves as though each monomer unit constitutes a separate and independent molecule in determining the thermodynamics of the solution, the concentration of monomer units at 100 mg/ml is only 2·27 M. It is clear that gross changes in the hydrodynamic properties of DNA are not observed in comparable concentrations of sucrose, ethanol, or cesium salts. A volume fraction of 0·1 does not approach the concentration required (about 0·5) for denaturation by an effective, water-miscible organic solvent at room temperature.

The negligible effect of PEO on water activity is indicated by direct measurements of vapour pressure, although under conditions somewhat different from the present experiments. In the lowest polymer concentration tested by Malcolm and Rowlinson (1957), a solution containing equal weights of water and PEO, molecular weight 5000, water activity in the solution was 0·974 of that of pure water at the same temperature, 64°C. Linear interpolation suggests a reduction of water activity by only 0·5% at a weight fraction of 0·1. Their vapour pressure measurements at this and higher concentrations imply a value of the interaction parameter of 0·4, which is in reasonably good agreement with that inferred from the second virial coefficient in the light scattering measurements of Alexandrowicz (1959). His value introduced into Flory's relations for the activity of the solvent, assuming 100 mg/ml PEO, MW 6800, gives a calculated relative water activity of 0·997, again hardly different from pure water.

It is clear that any dehydrating effect of the polymer is not more than that of the relatively modest salt concentration required for critical conditions.

A theoretical analysis of the role of salt—presumably its cations—in effecting the charge neutralisation necessary for close packing of DNA helices has not yet been attempted, but some features are evident. The large ratio between the critical concentrations of univalent and divalent ions, and the considerable specific variation in critical concentrations among singly charged ions indicate that site-binding, rather than ion atmosphere effects, should be inferred. In high PEO concentration, with low salt, $2\frac{1}{2}$–3 times as much ammonium chloride as sodium chloride is required. It may be noted that even in dilute solution, about half of the DNA phosphates hold closely associated cations (see Strauss *et al.*, 1967).

Some part of the variation of the critical polymer concentration as a function of salt may be attributable to a substantial Donnan effect between the bulk solution and the interior of the dense DNA phase.

Salts have a relatively minor effect on the properties of PEO at these concentrations, and there is no indication of cooperative changes. The intrinsic viscosity drops only 9·5% below its value in distilled water at the critical sodium chloride concentration for 100 mg/ml PEO, and the change remains linear in NaCl to at least 1 M (Bailey and Callard, 1959).

VI. Secondary and Tertiary Structure

Among the various physical properties that change under ψ conditions, the most readily amenable to simple kinetic measurement has been the circular dichroism. The molar ellipticity (in terms of nucleotide equivalents) changes from an extremum of -5×10^3 at 245 nm in dilute solution (and 0 at 260 nm) to about -2×10^5 at 260 nm, when measured as to minimize the effects of dichroic light scattering. However, the particles are sufficiently large as the transition approaches completion, even at the lowest practical DNA concentrations (as determined by the sensitivity, noise, and path length of present instruments), that the solution measurements are hard to interpret, and somewhat unreliable, except in the early part of the change. This in no way limits the detection of critical salt and polymer concentrations, and the early kinetics are of particular interest.

Solutions of T7 DNA, 2×10^{-5} N in nucleotides, reach a stationary condition under ψ conditions in about 40 hours, after which there are no appreciable CD changes or a slow decline attributable to continuing growth of the particles (Jordan *et al.*, 1972). Measurements at 48 hours at constant salt and varying PEO, or the converse, indicate essentially an all or none change across the interval of adjacent experimental concentrations, 4 mg/ml PEO or 0·005 M NaCl.

The early kinetics of the CD changes depend on PEO and salt concentration in a very intricate way (Jordan *et al.*, 1972; Lerman and Jordan, unpublished). There is a pronounced, extremely sharp optimum concentration for each, just slightly above critical, where the rate of change may be a hundredfold faster than at other over-critical concentrations. Beyond the optimum the dichroism proceeds to its final value approximately exponentially with time. In the narrow concentration zone between the critical and the optimum, the change begins very slowly and accelerates.

Both the optimum and the slow initial phase are characteristic of the kinetics of systems undergoing crystallisation, where the nucleation and growth of an orderly assembly are the rate-limiting steps.

Examination by X-ray scattering of the dense DNA phase collected from a dilute suspension provides two kinds of evidence with respect to its structure (Maniatis, 1971; Maniatis *et al.*, 1973). The scattering at large angles corresponds reasonably well with that expected for randomly oriented aggregates of locally ordered and parallel DNA helices with internal structure very close to the B fibre structure. The patterns clearly exclude any substantial contribution from the A or C fibre configurations. The region of well-ordered helicity in terms of the backbone appears to be greater than 10 but less than 100 nucleotide pairs, but very short range order between nucleotides is somewhat poorer. In general, the wide-angle scattering shows no changes as a function of salt and PEO in the positions of the maxima.

A peak of relatively strong scattering is observed at angles smaller than that of any intrahelix spacing. The angle is a simple function of the polymer concentration, and the breadth of the peak varies with both salt and PEO. It seems directly attributable to the interhelix spacing in an array of almost parallel helices. The greater breadth of the peak at high salt concentrations corresponds with the weaker electrostatic repulsion between helices, on which the orderliness of the array must depend. The inverse second-power of the spacing varies linearly with the PEO concentration; the helices become both closer and more orderly as the PEO increases. It is attractive to identify the array of parallel segments of helix with the ordered configuration implied by the nucleation kinetics observed with circular dichroism.

VII. Collapse of Single Molecules

The possibility of a unimolecular conformational change under the conditions of the expected phase transition has been investigated by measurement of the sedimentation rates of T4 and T7 DNA in suitable PEO and poly(vinyl pyrrolidinone) solutions (Lerman, 1971a). The dilution required for observation of unimolecular properties where aggregation is expected can be estimated from the comparison of the distances between molecules in solution, the dimensions of the ordinary extended form, and the rate at which collisions can be expected according to diffusion. The result indicates that transport properties can be observed with strong radioactive labelling. A small sample of solution containing ^{32}P labelled phage DNA was applied to the top of the contents of a centrifuge tube containing a uniform concentration of polymer and salt, together with a gradient of H_2O–D_2O to provide density stabilisation. Transport of the DNA that had occurred during centrifugation was determined in the usual way, by collecting the fractions after piercing the bottom of the tube, drying, and counting the samples. When the solutions contain 100 mg/ml PEO of nominal molecular weight 6000–7500, it is found that the sedimentation rate of DNA, corrected for the substantial viscosity of the solution, is approximately that expected on the basis of the usual sedimentation coefficient, if the total concentration of sodium present is 0·2 M or less. The width of the sedimenting zone remains about the same as that usually found for sedimentation of higher molecular weight DNA in sucrose gradients. When the sodium concentration exceeds 0·3 M, considerably more rapid sedimentation is observed, and at 0·4 M or higher sodium, DNA is found to sediment as an extremely broad, rapid zone. For T4 DNA the zone extends to about 600 S and shows a peak in the region of 300 S, while for T7 DNA the zone extends to about 400 S with a peak just above 200 S. The rapidly sedimenting distribution is not found at any salt concentration if PEO is present at only 40 mg/ml. Similar experiments using poly(vinyl pyrrolidinone) instead of PEO showed that rapid sedimentation could be obtained at high salt with 100 mg/ml of a polymer of nominal molecular weight 4×10^4 but not with the same concentration of nominal molecular weight 1×10^4. Rapid sedimentation is found with the latter at 250 mg/ml. Intermediate concentrations have not been examined.

In order to exclude aggregation as the basis of the rapid sedimentation, these experiments were carried out with sufficiently small samples of DNA that the sedimentation distributions were independent of the concentration within experimental error. Increasing values of the sedimentation coefficient were found when the amount of DNA contained in the 0·04 ml sample exceeded 10 picoequivalents of nucleotides (peq); however, the increment was only 9 % at 35 peq.

The supposition that individual T4 DNA molecules become rapidly sedimenting independently on entering the part of the solution containing both polymer and salt is supported by estimation of the second order rate constant for a hypothetical diffusion controlled dimerisation of the molecules. For T4 DNA, a nucleotide concentration of 10 peq/0·1 ml corresponds to a concentration of $0{\cdot}6 \times 10^{-12}$ M. The diffusion coefficient of T4 can be taken from the data of Reinert *et al.* (1971), or calculated from the sedimentation coefficient, as 3×10^{-9} cm^2/sec. If we assume that two DNA molecules will aggregate if they encounter each other in the polymer solution at a distance corresponding to the sum of their root-mean-square radii, about $1{\cdot}7 \times 10^{-4}$ cm, the rate constant as calculated by the rate of collision is $4{\cdot}2 \times 10^9$ M^{-1} sec^{-1}. Thus, dimerisation would proceed in water solution with a time constant of at least 400 sec. Since the polymer solution is four or five times more viscous than water, the time constant for dimerisation would be of the order of 30 min. Because of the spreading of the zone in the centrifuge tube, which dilutes the DNA, the actual rate should be even lower. The calculation shows that at the highest DNA concentration resulting in concentration-independent rates experimentally, there are an insufficient number of intermolecular collisions in the duration of the experiment to provide an appreciable fraction of dimers or larger aggregates.

An independent test for the absence of anomalous effects in the polymer solution as well as an additional demonstration that sedimentation measurements are unperturbed by aggregation is afforded by measurement of the sedimentation of whole T7 phage in the same family of gradients containing 100 mg/ml PEO and varying salt (Lerman, 1971a). The aggregation of T7 phage at a somewhat lower polymer concentration has provided a useful preparative method in which only low speed centrifugation is necessary (Yamamoto *et al.*, 1970).

In the experiments shown in Fig. 3 of Lerman (1971a), approximately 2×10^5 plaque-forming particles of T7 were sedimented in the standard PEO gradients. As compared with the experiments on T4 DNA, there are about seventy-fold fewer particles (neglecting the possibility that there are an appreciable number of nonviable phage particles present), a diffusion coefficient about twenty times higher (Dubin *et al.*, 1970) and a plausible collision radius about sixteen-fold smaller also. Accordingly, the rate of disappearance of individual phage, again assuming that every collision results in aggregation, would proceed with a long time constant, of the order of 1000 min, representing a larger safety factor than that allowed in the DNA experiments. The absence of anomalous effects due to the polymer or its combination with various salt concentrations is shown by the sedimentation of the phage in a narrow zone at very nearly the same sedimentation coefficient found in the absence of polymer, together with only slight, gradual changes as salt concentration is increased. Although aggregation would have been expected had the phage been allowed to stand in this medium for a long time, the properties of individual phage are observed, in agreement with the rate calculation.

Although the hydrodynamics indicates that each DNA molecule folds into a configuration offering up to ten times less frictional resistance, it cannot directly indicate internal structure of the compact state. However, some experiments on the inhibition of the ψ structure change by the binding of ethidium suggest that the modifications required for the transition affect only a small fraction of the total helical length. This conclusion would argue against models such as a supercoil for the compact state, in which the change would be distributed more or less uniformly over the entire length of the helix.

If ethidium is added in high concentration to PEO gradients such as those described in section 19, T4 DNA is found to remain in the extended, slowly sedimenting configuration (Lerman, 1971b). The effect appears to depend on enough ethidium binding to approach saturation of the intercalated mode. Since the association of ethidium with PEO reduces the effective ethidium concentration by about seven-fold (Lerman and Green, unpublished), a correspondingly higher ethidium concentration is required for saturation than would be necessary in polymer-free solution. These conditions approach the solubility limit of ethidium, and the clearest experiments have been carried out with somewhat lower than the standard PEO concentration, 80 mg/ml (see Fig. 1 of Lerman, 1971b). At substantially higher ethidium concentrations, the data suggest a tendency towards partial restoration of rapidly sedimenting forms. It is plausible to attribute this to the second mode of binding of ethidium to DNA, as stacked molecules external to the helix, displacing the intercalated mode. The failure of the heavily intercalated helix to undergo the changes necessary for the compact condition implies, conversely, that the parts of the molecule for which the structure is modified in the compact condition must be unable to accept intercalated ethidium, or the binding constant must be very severely reduced. The binding of ethidium to DNA was examined by means of the polarisation of its fluorescence and by the strong change in quantum yield on binding (Lerman and Green, unpublished) for a diverse series of salt and PEO concentrations. Since these measurements require only trace quantities of ethidium, no more than negligible changes in the transition parameters would be expected. The results showed no change in the binding equilibria as either the salt or polymer concentration crossed its critical value. It must be inferred that the binding properties are changed by the transition for only a small part of the DNA, if any. A simple folded structure in which roughly equal ordinary double helical segments lie bundled in a more or less parallel array, remains fully compatible with this inference. The structural features peculiar to the compact state would occur only at the points of folding between the straight segments, and these need not constitute more than a small fraction of the total number of nucleotide pairs.

It is interesting to compare the extraordinarily high sedimentation coefficient of the most rapidly sedimenting part of the distribution of T4 DNA under ψ conditions with the sedimentation coefficients that can be calculated according to a set of more or less plausible assumptions about the structure of the compact configuration. The results of a number of such calculations (Lerman, 1971a) show that relatively drastic changes are necessary in the hydrodynamic properties of DNA in order to account for the observed sedimentation rates. Although we are particularly concerned with the most rapid end of the distribution and cannot yet provide support for any hypothesis regarding the breadth of the distribution, it may be noted that the sedimentation patterns are reproducible and insensitive to DNA concentration.

A model based on the folding of T4 DNA to a bundle of straight parallel segments can be thought to generate a cylindrical particle of dimensions determined by the fold length and the spacing between adjacent helical segments. The diffraction results of Maniatis (1971) show that the spacing is relatively insensitive to salt concentration, while remaining moderately sensitive to the concentration of PEO. At 100 mg/ml PEO, the concentration used in the sedimentation experiments, there is a strong diffraction peak corresponding to a Bragg spacing of 3·82 nm. In the absence of additional lattice reflections or persuasive arguments for any other symmetry, it is convenient to assume that the helices are in a hexagonal array; the centre-to-centre distance between helices would then be 4·41 nm. Since this distance is about twice the diameter of the double helix, there is room for much water within the array but insufficient room for a randomcoil of PEO if its dimensions are unperturbed. Since the molecules are of uniform total length, the distance between folds determines the number of segments in the array as well as the length of the cylinder. We can suppose that the fold length may lie between the limits of 16 and 200 nm. The lower limit corresponds to the fold length determined for crystals of very low molecular weight DNA prepared by the slow cooling of DNA in alcohol–water mixtures after heating to the melting temperature (Giannoni *et al.*, 1969), and the upper limit corresponds to a fold length that can be invoked to explain the particles seen by Lang (1969) on the addition of alcohol to solutions of native T3 DNA. In any event, the fold length appears in the prevalent theoretical treatments of polymer crystallisation as a kinetically determined parameter which depends on the specific conditions of crystallisation, and is not uniquely determined by the properties of a particular polymer. Using Freifelder's (1971) estimate of the molecular weight of T4 DNA, which corresponds to a molecule of $1{\cdot}51 \times 10^5$ nucleotide pairs, we find these dimensions specify, in the limiting cases, an oblate cylinder with axial ratio of 16·2, or as the fold length increases, a progressively narrower cylinder, becoming prolate with an axial ratio of 2·67 at the other limit. With the shorter fold, the bundle would consist of $3{\cdot}14 \times 10^3$ rods, while with the longer it would have $2{\cdot}51 \times 10^2$, if a negligible additional length is required at each point of folding. Within the precision of this estimate, the cylinder can be replaced by an ellipsoid of revolution with the same axial ratio. We also assume that the partial specific volume of DNA remains the same as in the extended form. The sedimentation coefficient of the intermediate particle equivalent to a sphere is then 776s, a value somewhat greater than the highest values we have observed for T4ψ DNA. Using estimates of the frictional ratio for the oblate and prolate ellipsoids from the Perrin relation, the sedimentation coefficients for the poker chip and peanut models are then 469s and 713s respectively.† While it would be erroneous to impute too precise a significance to these values, it is clear that the folded chain can account satisfactorily for the rapid sedimentation. We may suppose that the breadth of the sedimentation distribution may be accountable in terms of imperfect folding leaving long, straggling loops or tails, as well as variation in fold length.

Two other models may be mentioned as representative of those that fail to account for sufficiently high sedimentation coefficients. If a T4 DNA molecule were to be folded in half and the two arms twisted into a very tight supercoil, it would constitute a stiff rod with about $1{\cdot}34 \times 10^3$ full turns. The helix axes are assumed to lie 2·9 nm from the axis of the supercoil. The axial ratio of the rod would be about 841, which by the Perrin relation for a prolate ellipsoid, specifies a frictional ratio of 10·8. The calculated sedimentation coefficient is then 94·6s, only about 50% higher than that of the extended, dilute solution form of T4 DNA.

Another approach to a more compact model without invoking any configuration change would be provided by a highly branched structure. Stockmayer and Fixman (1953) have provided a relation between the effective hydrodynamic radius of a randomcoil polymer and a particle of similar local configuration except that it consists of a number of uniform arms, all joined at a single point. Their relation suggests that a particle consisting of 100 segments of T4 DNA, each one 1/100th of the intact phage DNA, attached by one end to a common centre, would sediment slightly more than four times as fast as the unbranched molecule. The correction increases roughly as the square root of the number of branches when that number is large, so that a Medusa with about 400 snakes would have to be invoked to reach the required sedimentation velocity. Obviously the assumption of the normal configurational freedom on which the hydrodynamic expectation is based could not be fulfilled within so dense a cluster, and the calculation is not meaningful for an excessively large number of arms.

VIII. Folds in DNA

In the single crystals that have been grown as leaflets from solutions of a wide variety of linear polymers (see Keller, 1968) and including also low

† The difference between these values and the corresponding set offered previously (Lerman, 1971a), comes from the use of the experimental value of 3·82 nm for the Bragg interhelix spacing, rather than the previous, arbitrarily assumed value of 2·4 nm. The total volume of the DNA bundle with the present spacing is about 2·4 times as large as the interior volume of the T4 head. An interhelix spacing as close as that of the interior of the phage head can be obtained with calf thymus DNA, but a concentration slightly greater than 300 mg/ml PEO 6000 is required (Maniatis, 1971).

molecular weight DNA (Giannoni *et al.*, 1969), it has been demonstrated that the polymer chains lie folded into an essentially parallel array, essentially perpendicular to the broad faces of the crystal. There is considerable evidence (Mandelkern, 1970) that the continuity of the chains at the surface of the crystal, where folding must take place, is relatively disordered, rather than a neat, uniformly structured kink between adjacent segments. Thus, most of the crystal is an orderly array of parallel chains with less orderly loops of random length at both ends. In considering the applicability of this model to the compact state of DNA we may compare an estimate of the flexibility of DNA with the curvature required in the folded chain crystal. The bending between adjacent nucleotide pairs that can occur without drastic modification of the B structure can be estimated by application of the simple, worm-like chain theory, which relates the length of each monomeric residue, t, and the small allowed angle of bending between residues, β, to the rate of decline of the correlation between the directions of two parts of a stiff polymer chain as the distance between the two parts becomes increasingly great. The rate of decline is characterised by the persistence length, p, such that $p = t/(1 - \cos \beta)$. Estimates of the persistence length from hydrodynamic data are near 45·4 nm (Triebel *et al.*, 1971), which indicates a value of 7° for β, assuming the usual monomeric residue length of 0·34 nm. If it is assumed that there can be a coherent bending of an average of 7° between adjacent nucleotide pairs, the helix could curve with a radius of 2·8 nm. The closest distance between helix axes in ψ DNA indicated by our diffraction studies, assuming an hexagonal array, is 3·06 nm (in 300 mg/ml PEO), obviously considerably closer than double the allowed radius. Since it is unlikely in simple polymers that contiguous sections of the chain are adjacent in the crystal, such sharp bends are not necessary. In any event, that radius is small enough to almost accommodate joining next nearest neighbours, which would be about 5·1 nm apart in the dense lattice. The second nearest neighbours, at 7·6 nm in 100 mg/ml PEO, could be accommodated easily. Longer values of the persistence length have been proposed by other authors on the basis of light scattering measurements (Hays *et al.*, 1969; Eisenberg, 1969) which would require a bending angle of 5° and correspondingly, a radius of curvature of 3·9 nm. Although this curvature would allow slightly less proximity to contiguous segments, the argument is not fundamentally changed.

It may be supposed that crystals prepared from low molecular weight DNA by the procedure of Giannoni *et al.*—cooling partly denatured solutions in an alcohol–water mixture—are relatively more orderly, in that the denaturation step may permit sharper folding. Nevertheless, our observations on the circular dichroism spectra of crystals of sonicated calf thymus DNA (Jordan and Lerman, unpublished) are essentially the same as the CD spectra of calf thymus DNA under ψ conditions, without heat or alcohol.

The folded chain crystal has been proposed as a model for the configuration of DNA in phage heads and in eucaryotic chromosomes, under certain conditions (North and Rich, 1961; Rich, 1962).

Note added in proof

More recently study of packing and folding (Lerman, 1974) indicated that spontaneous ordering can prevail over macroscopic dimensions.

IX. Summary

In solutions of poly(ethylene oxide) and other non-complexing polymers, DNA adopts a compact molecular configuration, either as individual high molecular weight molecules or multi-molecular aggregates. The critical dependence of the transition on polymer and salt concentrations and on the polymer molecular weight supports the supposition that the change is due to competition for space and configurational freedom through repulsive, excluded volume interactions. DNA retains predominately a structure close to the B fibre structure. The properties of the compact state are in accord with the folded chain structure that is characteristic of crystals of most linear polymers, but alternative models in which there is substantial parallelism of adjacent helices are not excluded. The capacity for chain folding seems to be a simple property of the DNA molecule, not dependent on the strong external constraints or complex formation.

References

Albertsson, P.-Å. (1971). "Partition of Cell Particles and Macromolecules" (2nd ed.). Wiley, New York.

Alexandrowicz, A. (1959). *J. appl. Poly. Sci.* **40**, 107.

Bailey, F. E. jr. and Callard, R. W. (1959). *J. appl. Poly. Sci.* **1**, 56.

Dubin, S. B., Benedek, G. B., Bancroft, F. C. and Freifelder, D. (1970). *J. molec. Biol.* **54**, 547.

Eisenberg, H. (1969). *Bipolymers*, **8**, 545.

Flory, P. J. (1953). "Principles of Polymer Chemistry". Cornell University Press, Ithaca, New York.

Freifelder, D. (1970). *J. molec. Biol.* **54**, 567.

Giannoni, G., Padden, F. J. and Keith, H. D. (1969). *Proc. natn. Acad. Sci. U.S.A.* **62**, 964.

Hays, J. B., Magar, M. E. and Zimm, B. H. (1969). *Biopolymers* **8**, 531.

Jordan, C. F., Lerman, L. S. and Venable, J. H. jr. (1972). *Nat. new Bio.* **236**, 67.

Keller, A. (1968). *Rep. Prog. Physics*, **31**, 623.

Lang, D. (1969). *J. molec. Biol.* **46**, 209.

Laurent, T. C. and Ogston, A. G. (1963). *Biochem. J.* **89**, 249.

Lerman, L. S., (1971a). *Proc. natn. Acad. Sci. U.S.A.* **68**, 1886–90.
Lerman, L. S. (1971b). *Progress in molec. subcell. Biol.* **2**, 382–391.
Lerman, L. S. (1974). *Cold Spring Harb. Symp. quant. Biol.* **XXXVIII**, (in press).
Malcolm, G. N. and Rowlinson, J. S. (1957). *Trans. Faraday Soc.* **53**, 921.
Mandelkern, L. (1970). *Prog. Polym. Sci.* **2**, 163.
Maniatis, T. P. (1971). "Ph.D. Dissertation", Vanderbilt University.
Maniatis, T. P., Venable, J. H., jr. and Lerman, L. S. (1973). *J. molec. Biol.* (in press).
North, A. T. C. and Rich, A. (1961). *Nature, Lond.* **191**, 1242.
Ogston, A. G. (1962). *Archs biochem. Biophys. Suppl.* **39**.
Okubo, T. and Ise, N. (1969). *J. Phys. Chem.* **73**, 1488.
Reinert, K. E., Strassburger, J. and Triebel, H. (1971). *Biopolymers* **10**, 285.
Rich, A. (1962). *Pont. Acad. Scient. Scripta Varia,* **22**, 1.
Stockmayer, W. H. and Fixman, M. (1953). *Ann. N.Y. Acad. Sci.* **57**, 334.
Strauss, V. P., Helfgott, C. and Pink, H. (1967). *J. phys. Chem.* **71**, 2550.
Triebel, H., Reinert, K. E. and Strassburger, J. (1971). *Biopolymers* **10**, 2619.
Yamamoto, K., Alberto, B., Benzinger, R., Lawhorne, L. and Trieber, G. (1970). *Virology* **40**, 734.

CHAPTER 20

The Interaction of Nucleotides with Bovine Pancreatic Ribonuclease

F. W. BENZ AND G. C. K. ROBERTS

National Institute for Medical Research, Mill Hill, London N.W.7

I. Introduction

Bovine pancreatic ribonuclease catalyses the hydrolysis of the phosphodiester bonds of single-stranded RNA in a two-step reaction. The first step is a transphosphorylation to give an oligonucleotide terminating in a 2′ : 3′ cyclic phosphate, which is then hydrolysed to a terminal 3′ phosphate (Brown and Todd, 1952; Brown *et al.*, 1952; Markham and Smith, 1952). The enzyme is specific for phosphodiester bonds in which the 3′-linked nucleotide is a pyrimidine, and pyrimidine dinucleotides (e.g. CpC, CpA) or 2′ : 3′-cyclic phosphates (C > p and U > p) are substrates. The isomeric pyrimidine nucleoside monophosphates (e.g. cytidine 2′-, 3′- or 5′-monophosphates) are competitive inhibitors; since the 3′-monophosphate is the product of the enzyme-catalysed hydrolysis of the cyclic phosphate intermediate, the enzyme shows quite marked product inhibition. Although the major specificity is for a 3′-linked pyrimidine nucleotide, the nature of the 5′-linked nucleotide can make a very considerable difference to the rate of the transphosphorylation.

This enzyme and its interaction with nucleotides has been studied intensively for over thirty years, and with the recent crystallographic work of Kartha *et al.* (1967) and Wyckoff *et al.* (1967), much of the considerable body of knowledge which has been accumulated in this period can be rationalised on the basis of the three-dimensional structure of the protein. The excellent review by Richards and Wyckoff (1971) contains a discussion of most of the available information on the interaction of nucleotides with the enzyme and on the mechanism of the enzymatic catalysis in this context. In the present review we shall not attempt to duplicate their comprehensive coverage of the literature; rather we shall illustrate the contributions of some of the many techniques which have been applied to ribonuclease, and attempt to point out remaining areas of uncertainty in our knowledge of the ribonuclease–nucleotide system, and to indicate fruitful directions for future work.

In order to make the subsequent discussion intelligible, we begin with a description of the structure of the enzyme–nucleotide complexes as determined by X-ray diffraction, and then consider how the interactions revealed in this way explain the specificity of the enzyme. Since many of the fine details of the functioning of ribonuclease will require spectroscopic techniques for their elucidation, we shall also discuss briefly the correlation between spectroscopic and crystallographic studies of the enzyme–nucleotide complex. We shall then discuss the limited information available about possible changes in protein conformation on nucleotide binding, and consider why these may not be observable in the crystal; finally we shall point out some of the major gaps in our understanding of this system. Throughout, the emphasis will be on the interpretation of experiments carried out in solution in terms of the structure in the crystalline state. It seems clear that one is now justified in believing that proteins in solution have structures which are substantially the same as those in the crystal. Ribonuclease is indeed enzymatically active in the crystalline state (Doscher and Richards, 1963; Bello and Nowoswiat, 1965), and also preserves the characteristic reactivity of histidine residues 12 and 119 towards bromoacetate (Bello and Nowoswiat, 1965). However, small differences in conformation, and particularly in conformational flexibility, between crystal and solution are to be expected, and it is possible that these could have major functional consequences. What evidence there is for differences in the conformation of ribonuclease between these two states will be discussed in the following sections (particularly V and VI).

II. Description of the Ribonuclease–Nucleotide Complex as Determined by X-ray Crystallography and its Implications for Specificity

A. THE STRUCTURE OF THE COMPLEX

At the present time, 2 Å electron density maps are available for both ribonuclease A (Kartha *et al.*, 1967) and ribonuclease S (Wyckoff *et al.*, 1970). Ribonuclease S is produced from ribonuclease A by the action of subtilisin, which cleaves the Ala 20–Ser 21 peptide bond (Richards and Vithayathil, 1959). Some preparations of subtilisin also lead to cleavage of the Ser 21–22 bond (Doscher and Hirs, 1967; Gross and Witkop, 1967), but this seems to make no difference to the overall properties of ribonuclease S, which has closely similar catalytic activity to the parent molecule, ribonuclease A. While a list of the coordinates of all non-hydrogen atoms in the crystal struc-

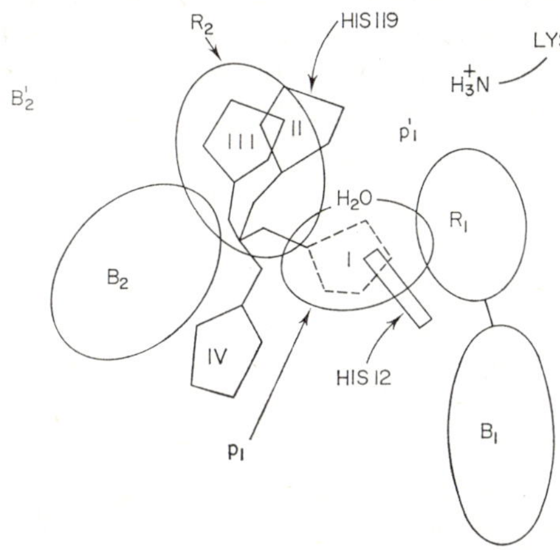

FIG. 1. A schematic view of the active centre as deduced from the X-ray data from the protein and several substrate related complexes. B_1, R_1, p_1, R_2 and B_2 indicate the relative positions of the bases, riboses and phosphate of the dinucleotide analogue UpcA. Position p_1 is occupied by $SO_4^=$ in the protein crystal. 3′-CMP, UMP and analogues of these occupy B_1, R_1 and p_1 predominantly. 5′-AMP occupies B_2 and R_2 and p_1 while 3′-AMP and 3′, 5′A>p occupy B_2 and R_2 predominantly, and possibly to a lesser extent B_1 and R_1. B_2' is the probable position of the second pyrimidine in dinucleotides such as CpU. The phosphate position in C>p cannot be observed owing to digestion but would be at p_1' if the base occupies the same position as in 3′-CMP. Four His 119 positions are indicated. I coincides with p_1 but is a possible position in the absence of $SO_4^=$ or nucleotides. II is behind III and may be occupied by solvent. III is slightly stabilised by 3′-CMP. IV is the position occupied when B_2 and R_2 are occupied by adenine phosphates. His 12 is behind p_1 and R_1. There is a solvent molecule, presumably water, behind p_1 as indicated by H_2O. Lys 41 enters from the upper right and is not in contact with p_1, but might contact p_1'. Asp 121 enters from below and may contact His 119 in position IV. (From Richards and Wyckoff, 1971.)

ture of ribonuclease S has been published (Wyckoff *et al.*, 1970), a corresponding list for ribonuclease A has not yet appeared, so a detailed comparison of the structures is not possible. However, Dickerson and Geis (1969) have prepared a drawing comparing the main-chain conformations, and apart from the expected differences in the region of the subtilisin cleavage, the structures are clearly very similar. In the absence of detailed information for ribonuclease A, the following description of the active site and of the binding of nucleotides will be based solely on studies of ribonuclease S (Wyckoff *et al.*, 1970; Richards *et al.*, 1970, 1971; Richards and Wyckoff, 1971), and the assumption will be made that at least this region of the molecule is identical in ribonuclease A.

A schematic diagram of the active site of ribonuclease and the positions occupied by mono- or dinucleoside phosphates when bound to this site is shown in Fig. 1. As was mentioned in the introduction, the enzyme is quite specific in its endonuclease activity in that it cleaves primarily on the 3′ end of pyrimidine bases. Site B_1 seems to be the major specificity site for binding of pyrimidine bases. This can be appreciated more fully with the aid of Fig. 2,

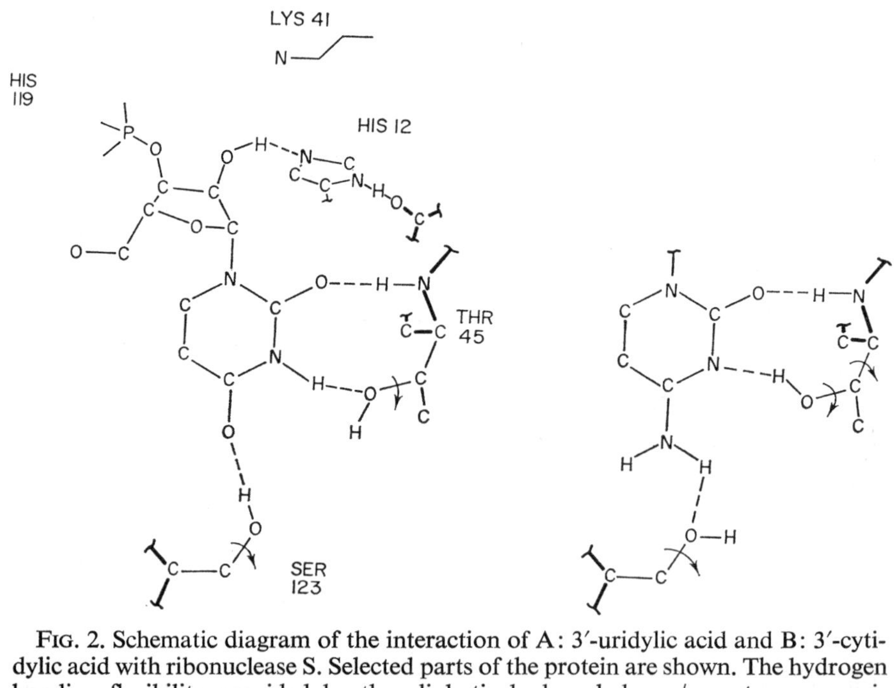

FIG. 2. Schematic diagram of the interaction of A: 3′-uridylic acid and B: 3′-cytidylic acid with ribonuclease S. Selected parts of the protein are shown. The hydrogen bonding flexibility provided by the aliphatic hydroxyl donor/acceptor groups is shown, together with the fixed requirements of the chain NH group. For further details see text. (From Richards *et al.*, 1970.)

which is a diagrammatic representation of the active site region of RNase S with one molecule of 3′-cytidine monophosphate (3′-CMP) or 3′-uridine monophosphate (3′-UMP) bound to it. It can be seen from this figure that the specificity of RNase can be satisfactorily accounted for by the formation of three hydrogen bonds between the pyrimidine base and the enzyme. Both cytidine and uridine have a carbonyl group in position 2 which seems from the X-ray difference Fourier to form a hydrogen bond with the backbone NH of threonine 45. Since the atoms in the 3 and 4 positions on cytidine and uridine differ, the enzyme must be able to accommodate both. This is accomplished by using the aliphatic hydroxyl groups of threonine 45 and serine 123 which can function either as donors or as acceptors of the hydrogen bonds, depending on the requirements of the base. A stereodiagram of 3′CMP bound to the active site of RNase and hydrogen bonded according to the above scheme is shown in Fig. 3. Binding to site B_1, analogous to that shown for 3′-CMP, has been observed in crystals with 3′-UMP, 2′-CMP, and 4-thio-UMP. Purines do not as a rule bind to site B_1, but at very high concentrations some binding of purine nucleotides (e.g. 3′-AMP) to site B_1 has been observed; however the hydrogen bonding scheme for purine binding, if any exists, is not yet apparent. Site R_1 in Fig. 1 refers to the position of the ribose ring attached to the position 1 of the pyrimidine base found in site B_1. Interactions between the ribose in site R_1 and the enzyme seem to be less well defined. In Fig. 2 the 2′OH group of the ribose is shown to be hydrogen bonded to the N-3 of histidine 12, whose N-1 is in turn hydrogen bonded to the backbone carbonyl of threonine 45. This geometry is thought to be that appropriate to the initial stages of the transphosphorylation reaction and is discussed later in the section on the catalytic mechanism. When on the other hand 2′- or 3′-nucleoside monophosphates are bound to the active site, a protonated histidine 12 may interact with the phosphate mono or dianion (see below, p. 109) in which case the 2′OH group of the ribose may be free to hydrogen bond to the side chain carbonyl of asparagine 44.

Site p_1 in Fig. 1 is the main anion binding site of RNase. The bound phosphate groups of 3′-CMP and 3′-UMP are found in this position. In addition the 5′-phosphate groups of certain purine mononucleotides are also observed to bind there (e.g. 5′-AMP); however the phosphate group of 3′-purine mononucleotides does not interact with RNase at the p_1 site. The phosphate monoanion of an oligonucleotide substrate for RNase is also believed to bind at site p_1, by inference from the binding properties of a pseudosubstrate UpcA which is discussed below. On the other hand, if the positions of the pyrimidine base in site B_1 and the ribose ring in site R_1 remain relatively fixed, then the position of the phosphate monoanion in a 2′ : 3′-nucleoside cyclic phosphate will probably be closer to site p_1' which according to Fig. 1 is somewhat closer to the terminal amino group of lysine 41.

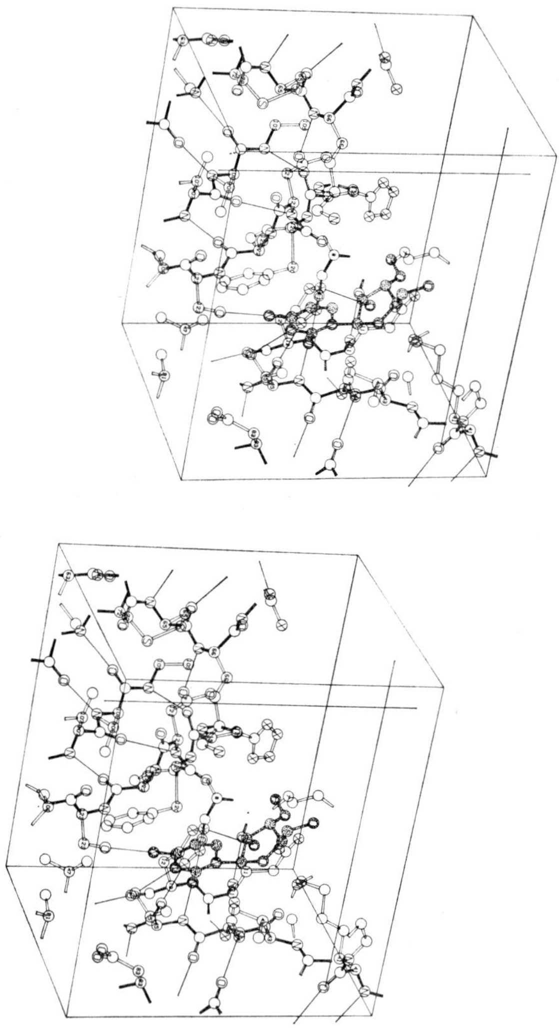

FIG. 3. Stereodiagram of the binding of 3′-CMP to ribonuclease S. The 3′-CMP molecule is stippled to distinguish it clearly from the protein. (From Richards *et al.*, 1970.)

Sites B_2 and R_2 in Fig. 1 are the binding sites for the purine base (B_2) and ribose ring (R_2) on the 5′ end of the 3′, 5′-dinucleoside phosphate or oligonucleotide undergoing hydrolysis. The B_2 site should perhaps at this stage be considered an adenine binding site since most of the evidence concerning it has been obtained with mononucleotides containing adenine (e.g. 5′-AMP, 3′-AMP). However, in addition to these purine mononucleotides, the binding of a very interesting dinucleoside phosphonate, UpcA, has been studied. This molecule differs from the dinucleoside phosphate UpA, which is a substrate for RNase, in that the 5′ oxygen of adenosine has been replaced by a methylene ($-CH_2-$) group. This alteration renders the molecule resistant to ribonuclease hydrolysis and thus its binding can be studied by X-ray crystallography. A stereodiagram of the complex of UpcA and ribonuclease is shown in Fig. 4. At this time of writing no detailed description of the interactions between the adenine ring in site B_1 and the enzyme has been published; however glutamic acid 111, asparagine 71 and perhaps glutamine 69 would seem to be likely candidates for interaction with the adenine ring.

No crystallographic studies of a dinucleoside phosphonate with a pyrimidine base on the 5′ end have been reported, so no information is yet available on the binding of a second pyrimidine base to the enzyme from this source. However, at high concentrations of pyrimidine mononucleotides an increase in electron density appears at site B_2' and this may represent a possible site for the second pyrimidine base in a pyrimidine 3′,5′-pyrimidine dinucleoside phosphate substrate. Little or no increase in electron density appears at B_2, and thus there is no evidence that site B_2 binds pyrimidines to any extent.

In addition to the binding sites for the substrates and mononucleotide inhibitors of RNase, Fig. 1 also illustrates the relative immobility of histidine 12 in contrast to the substantial mobility of histidine 119 in various complexes. The exact position of histidine 119 in the crystal structure seems to be variable in that electron density may appear and disappear from a number of locations labelled I, II, III, IV in Fig. 1. Some positions seem to be stabilised when certain small molecules are complexed, for example position III is stabilised when 3′-CMP is bound to the enzyme. This can be seen quite clearly in the stereodiagram of the 3′-CMP-RNase S complex (Fig. 3). In this complex the phosphate group is twisted away from lysine 41 and sits above the backbone NH of phenylalanine 120 between the N-3 of histidine 12 and presumably the N-1 of histidine 119 in position III. As suggested before, in this complex the 2′OH of the ribose may hydrogen bond to the side chain asparagine 44. Lysine 41 is somewhat above and to the right of the phosphate in the X-ray map of this complex. When UpcA is bound (Fig. 4) histidine 119 is displaced from position III by the ribose of the adenosine portion of the phosphonate and is observed instead to be stabilized in position IV. As seen from the stereodiagram, the adenine ring is somewhat above histidine 119 when in position

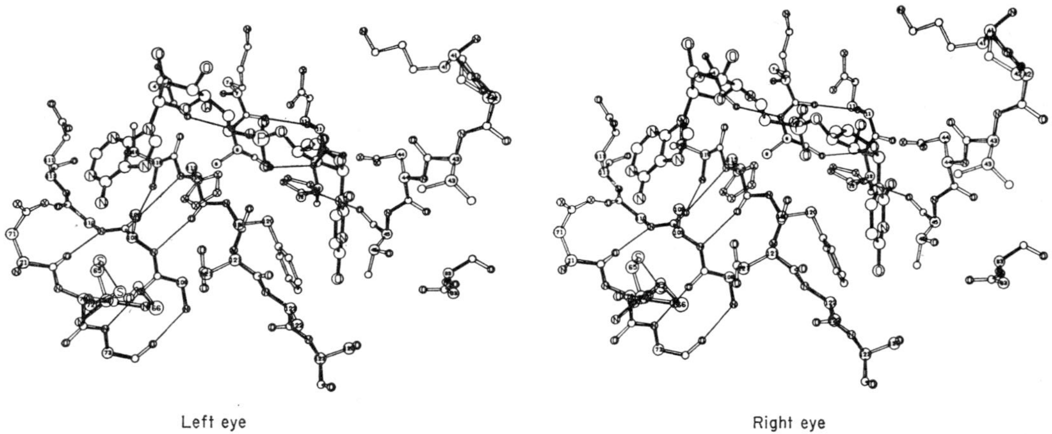

FIG. 4. Stereodrawings of the dinucleotide phosphonate UpcA bound to RNase-S. Histidine 119 is shown in position IV where it is forced to by the adenosine. Lys 41 is in the position found in the DNP-Lys 41 derivative. In the native protein it is lower and closer to the phosphate position, but it is not long enough to contact the phosphate. The UpcA is an interpretation of the 2 Å resolution electron density difference map and the CH_2 which is bound to the phosphorus is crowding His 119. (From Richards and Wyckoff, 1971.)

IV and indeed there may be a partial stacking interaction. Aspartic acid 121 is very close to histidine 119 in this position and may hydrogen bond with it. The geometry of this UpcA complex is virtually identical to that observed with simultaneous binding of 3′-CMP and 3′-AMP. Close inspection of the UpcA stereodiagram shows that histidine 119 is very close to the 5′-methylene group of the adenosine in the UpcA; this is of some importance in considering the catalytic mechanism, and will be referred to in section III.

The remaining two positions where electron density is found for histidine 119 are labelled I and II. Position I is normally the position occupied by the sulphate anion in crystals of RNase S which are grown in strong ammonium sulphate solutions, but it could be occupied by histidine 119 in the absence of sulphate or phosphate. Although histidine 12 in its normal constrained location and histidine 119 in position II are not coplanar, they are sufficiently close to one another to interact. This proximity should not imply, however, that such interaction between the two active site histidines must exist. Position II appears to be the least well characterised of the possible locations for histidine 119, and may represent a molecule of solvent rather than a histidine ring.

The only other amino acid side chain depicted in Fig. 1 is that of lysine 41. As illustrated in the figure the terminal ammonium group cannot quite reach the main anion binding site p_1 but it could tip down quite close to p_1'. Since its neighbouring residue cysteine 45 is in a disulphide bridge, lateral motion of the lysine 41 residue towards the p_1 site would seem to be ruled out unless a substantial movement of the half of the ribonuclease molecule containing lysine 41 towards the p_1 site occurs; this possibility will be discussed in section V. In 2′ : 3′-cytidine cyclic phosphate the phosphate may in fact be much closer to lysine 41 in site p'_1 and contact could be effected without the necessity for a conformational change. If histidine 119 were to be somewhat between positions I and IV, interactions between it and both the phosphate monoanion of the cyclic phosphate and aspartic acid 121 would still be possible.

Other residues in the immediate vicinity of the nucleotide binding site are best illustrated in Fig. 4. Below the ribose ring in position R_1 is a "hydrophobic core" made up of twelve or more non-polar amino acids. Behind the pyrimidine ring occupying site B_1 is found the aromatic side chain of phenylalanine 120; as is apparent from the figure these two rings do not appear to be coplanar. In front of the uridine ring lies the side chain of valine 43. Thus the pyrimidine base must slide into the active site between these two side chains. It should be noticed that in the bound state the 6-position and especially the 5-position on the pyrimidine base are not surrounded bv the protein but point out somewhat into the solvent.

B. IMPLICATIONS FOR THE SPECIFICITY OF THE ENZYME

With the model of the active site geometry of RNase S and the specific interactions between this site and the various nucleotides discussed above, one has sufficient information to provide an educated guess as to whether a particular dinucleoside phosphate, nucleoside cyclic phosphate or mono-nucleotide will bind to RNase and whether it will serve as either a substrate or inhibitor. It seems clear that a major force of attraction between a nucleotide and ribonuclease is provided by the ionic interaction between the cationic cluster of amino acids (Lys 7, His 12, His 119, Lys 41, Arg 39, etc.) at the active site and the anionic ligand. Thus simple anions (Winstead and Wold, 1965; Irie, 1965) or anionic polymers (Anfinsen and White, 1961; Scheraga and Rupley, 1962; Mora, 1962) are effective inhibitors of the enzyme. That the base and ribose ring, binding in sites B_1 and R_1, actually do participate in the binding process was demonstrated by Ukita *et al.* (1961b). These authors examined a variety of derivatives of 2(3H)pyrimidinone including the natural pyrimidine bases and found that these bases were themselves weak inhibitors of the enzyme. This binding strength was increased by conversion of the base to its nucleoside and a more dramatic increase resulted on subsequent conversion to the nucleotide substantiating the importance of the anion binding site stressed above. However, 2′(3′)-UMP was a much stronger inhibitor than the α, β methyl-D-ribofuranoside 2′(3′)-phosphate, further illustrating the importance of the interactions between the base and the enzyme. In addition, the inhibition due to anions is reduced at high ionic strength, whereas 2′-CMP binding is little affected, suggesting that the cytosine moiety contributes substantially to the binding strength (Nelson *et al.*, 1962).

With respect to the interactions of the base in site B_1, many of the findings concerning specificity can be rationalised using the X-ray model as a guide. Thus it is usually found that a good substrate will have a carbonyl function on the 2 position on the pyrimidine ring (Gassen and Witzel, 1967). If this is replaced by a thionyl group activity falls somewhat and substitution with an amino group renders the molecule inert to ribonuclease (Holy and Bald, 1971). The 3 position will tolerate a hydrogen (as found in uracil) but replacement of this with a methyl or N-oxide group yields a molecule which is not a substrate (Letters and Michelson, 1962; Gassen and Witzel, 1967; Witzel, 1963). In addition Gilham (1962) observed that substitution of a bulky group in the 3 position of uracil rendered U*pC resistant to RNase hydrolysis, whereas CpU* was hydrolysed rapidly. This can be readily accounted for by the X-ray model since the B_1 site seems to be more restrictive than does the B_2 site.

Variations in the 4 position of the pyrimidine ring appear to be tolerated better than those on the 2 and 3 position, as might be expected from the rather weak interaction between the functional group in position 4 and serine 123 indicated in the X-ray structure. Ukita *et al.* (1961b) found that 4-deoxyuridine was an inhibitor, though a weaker one than uridine. Bulky substituents on the 4-amino group of cytidine cyclic phosphate do not, as a rule, eliminate activity as substrates (Gassen and Witzel, 1967; Witzel, 1963), although substantial changes in K_m or k_s may result. Cleavage after the 4-thio-uracil base was found to occur in tobacco mosaic virus RNA (Mandel *et al.*, 1957) and similarly Gassen and Witzel (1967) observed the 2′ : 3′-cyclic phosphate of 4-thiouridine was split to 4-thio-UMP by ribonuclease. Even replacement of the 4 carbonyl group on uracil by methyl does not destroy activity (Bald and Holy, 1971). Extensive substitutions on the 5 position are tolerated. The X-ray model shows clearly that the 5 position of the base does not come in contact with the enzyme, but points out into the solvent. Ukita *et al.* (1961b) found that uracil substituted in the 5 position with halo, cyano, nitro or methyl groups could still inhibit ribonuclease hydrolysis of C>p. Modifications were made by Holy and Bald (1971) on the corresponding cyclic phosphate and all compounds were substrates. Polyuridylic acid or polycytidylic acid halogenated in the 5 position with chlorine, bromine or iodine serve as substrates for the enzyme (Letters and Michelson, 1962; Massoulie *et al.*, 1966) and so do the corresponding cyclic phosphates (Gassen and Witzel, 1967; Barker *et al.*, 1961). Davis *et al.* (1959) observed that ribonuclease was able to cleave at the 3′ side of ribothymidine bases of yeast RNA and also to hydrolyse the resulting cyclic phosphate. Wigler and Choi (1964) synthesised the 3′ sulphate of 5-fluorodeoxyuridine and found it to competitively inhibit RNase. Pseudouridine cyclic phosphate is a substrate as is its 1 methyl derivative. However, as would be expected, 1,3 dimethyl pseudouridine cyclic phosphate is not (Gassen and Witzel, 1967).

Few variations have been tested with regard to position 6 on uracil. 6-Methyl U>p is not a substrate but when oxidised to 6-carboxy U>p activity apparently returns (Holy and Bald, 1971). Moving the 4 carbonyl group of uracil to the 6 position decreases activity substantially (Bald and Holy, 1971). These latter substitutions only affect binding if the B_1 site is being considered. Linking the modified nucleosides on the 5′ end of a uridine 3′5′ diester has little effect on the transphosphorylation reaction (Holy *et al.*, 1972). Nucleotides containing 5,6-dihydrouridine are good substrates (Gassen and Witzel, 1967).

Before detailed X-ray data was available, a number of experiments shed some light on the stereochemical relationship between the base and sugar ring required for ribonuclease hydrolysis. Gassen and Witzel (1965) found that the cyclic phosphate of α-cytidylic acid would not serve as a substrate nor would it inhibit the hydrolysis of a normal substrate; however the α-cytidine 2′ or 3′ monophosphate were found to be inhibitors. β-D-lyxo-uridine cyclic phosphate (Fig. 5) in which the cyclic ester grouping is on the same side of the ribose ring as the base and 5′ carbon, as opposed to the

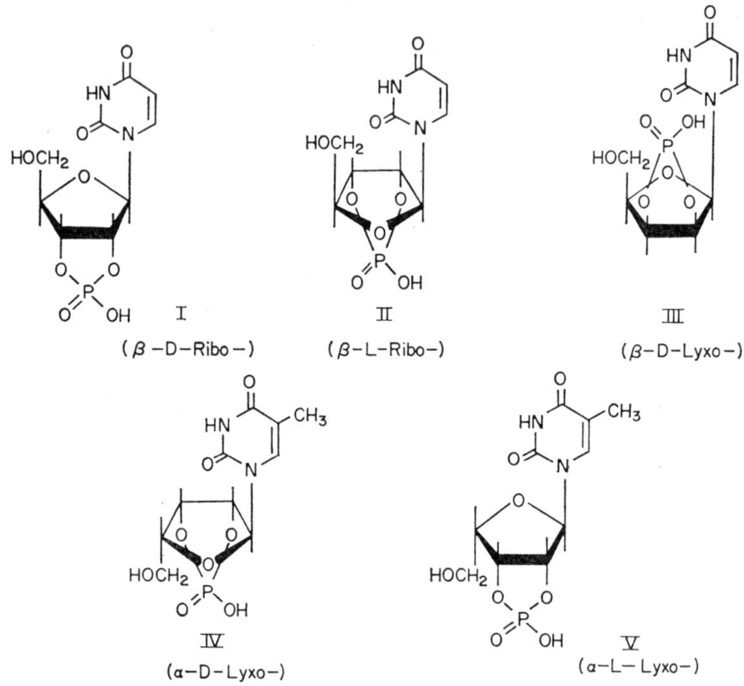

FIG. 5. Structures of nucleotide cyclic phosphates containing different pentose sugars which were tested as substrates for ribonuclease. (From Holy and Sörm, 1968.)

normal β-D-ribo configuration found in RNA, was shown by Ukita *et al.* (1961a) to be inert to the actions of RNase. The 2′(3′) or 5′ phosphate monoesters of this nucleoside were inhibitors but their binding strengths were approximately equivalent to that of the simple orthophosphate ion (Ukita *et al.*, 1961b), thus making any contribution to binding by the nucleoside portion unlikely. The monoesters, which are potentially dianionic, were in fact more active as inhibitors than the 2′ : 3′-cyclic diester monoanion, suggesting that the dianionic nature of these molecules is the controlling factor in binding; this is probably the case for the α2′(3′)CMP inhibitor as well.

Holy and Sörm (1968) prepared α-L-lyxo- and α-D-lyxo-uridine 2′ : 3′-cyclic phosphate and observed that while the α-L-lyxo molecule was a substrate for both RNase A and S, the α-D-lyxo was not nor did it inhibit. These authors also subsequently showed that the β-L-ribo and β-L-lyxo analogues of uridine 2′ : 3′-cyclic phosphate are not substrates (Holy and Sörm, 1969a, b). The unimportance of the geometry of the 5′ carbon atom for actiavity was already demonstrated above since the α-L-lyxo-uridine>p is substrate and it appears that halo substituents in the 5′ position of normal β-D-ribofuranoside are also tolerated (Michelson, 1962) as is the 5′-deoxy modification (Michelson, 1963). The effect or lack of effect of all the above modifications on activity or binding can in general be correlated with the model of binding based on X-ray crystallography. Evidence for the importance of the 2′OH group on the sugar for substrate activity is derived mainly from mechanistic concepts and is discussed in that section. However, Ukita *et al.* (1961b) did observe that the inhibiting activity of thymine ribofuranoside and cytosine ribofuranoside was decreased when the sugar ring was changed to the deoxy ribofuranosyl group.

In this section we have not discussed those nucleotides whose activity is not readily explicable, as yet, from an examination of the crystal structure. A discussion of these problem molecules is postponed to Section VI.

III. The Catalytic Mechanism

The structure of the complex between ribonuclease and the substrate analogue UpcA as determined by X-ray crystallography (illustrated in Fig. 4), clearly indicates the involvement of histidine residues 12 and 119, and some of the extensive chemical and spectroscopic evidence implicating these residues in substrate and inhibitor binding and catalysis will be discussed in Section IV.

In the last twelve years a considerable number of more or less closely related mechanisms have been proposed to describe the role of these two histidines in catalysis (Findlay *et al.*, 1962; Witzel, 1963; Wang, 1968; Hammes, 1968; Usher, 1969; Roberts *et al.*, 1969). These mechanisms have been described and discussed in detail by Richards and Wyckoff (1971), and the present discussion will therefore be limited to a consideration of the new evidence which has been obtained recently by Usher, Eckstein and their colleagues (Usher, *et al.*, 1970; Usher *et al.*, 1972; Eckstein *et al.*, 1972).

The hydrolysis of phosphate esters proceeds through a pentacovalent

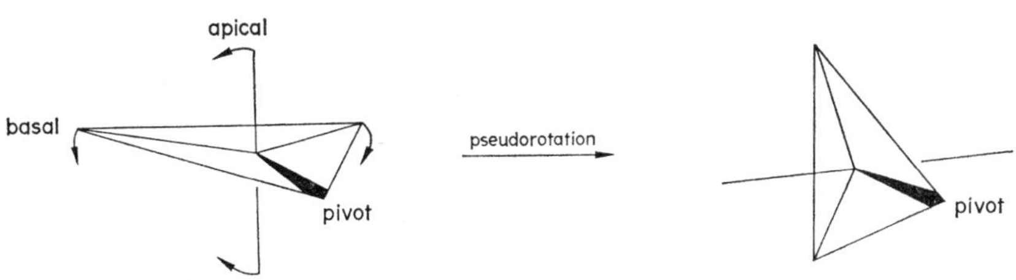

FIG. 6. The process of pseudorotation of pentacovalent phosphorus intermediates. The difference between apical and basal (or equatorial) substituents is also illustrated. (From Usher, 1969.)

intermediate which has the geometry of a trigonal bipyramid (for a review, see Westheimer, 1968). In this intermediate, three of the substituents on the phosphorus atom form an equatorial plane, with angles of 120° between them, and the remaining two substituents are apical, at right angles to this plane (see Fig. 6). The equatorial and apical substituents can interchange by pseudorotation of the bipyramid, as shown in the figure. There are a number of constraints on the arrangement of substituents in these intermediates, which can be summarised as follows:

(i) When there is a five-membered ring, as in the 2′ : 3′-cyclic phosphates, this spans one equatorial and one apical position, to avoid excessive ring strain.

(ii) Attacking and leaving groups occupy apical positions.

(iii) The preference for apical positions is in the order $OR > O^- > CH_2$.

Usher (1969) has pointed out that these rules allow one to classify all proposed mechanisms for ribonuclease action as "in-line" or "adjacent" mechanisms. The difference between these two kinds of mechanism is illustrated in Fig. 7. In an in-line mechanism (for example, Findlay *et al.*, 1962; Roberts *et al.*, 1969), the attacking group enters in such a way that the first-formed intermediate has the leaving group already in the other apical position. For this kind of mechanism, separate groups on the enzyme must be involved in deprotonating the attacking group and in protonating the leaving group.

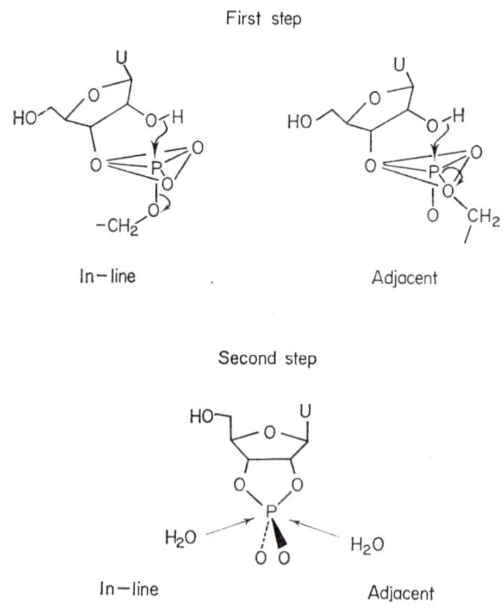

FIG. 7. The in-line and adjacent mechanisms for the two steps of ribonuclease action. In the adjacent mechanisms, the molecule would undergo a pseudorotation before product formation. (From Usher, 1969.)

In an adjacent mechanism (for example, Wang, 1968; Hammes, 1968), the leaving group is in an equatorial position in the first-formed intermediate, so that a pseudorotation is needed to bring it into an apical position. In adjacent mechanisms, the same group on the enzyme can both deprotonate the attacking group and protonate the leaving group.

The characteristics of these two pathways are discussed by Usher (1969), Roberts *et al.* (1969) and Richards and Wyckoff (1971). Roberts *et al.* (1969) argued that the structure of the enzyme-substrate complex as deduced from crystallographic (see Richards and Wyckoff, 1971) and n.m.r. (Meadows and Jardetzky, 1969; Meadows *et al.*, 1969) studies strongly suggests that an in-line mechanism holds. As can be seen from Fig. 4, the disposition of the attacking 2′O and the leaving 5′O in the complex is such that a pentacovalent intermediate with both these groups apical could form readily. The details of the

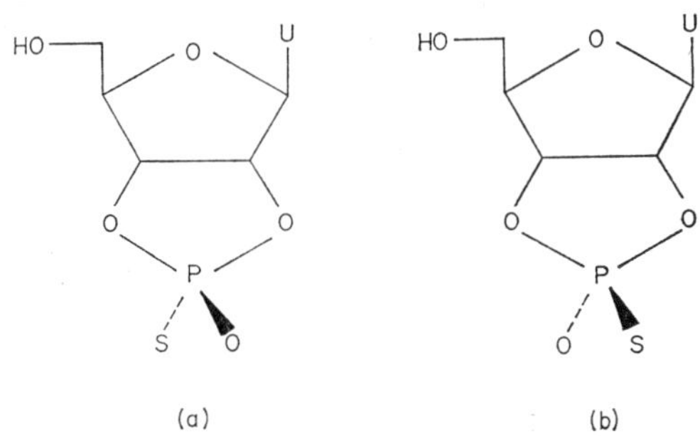

FIG. 8. The two diastereoisomers of uridine 2′ : 3′-cyclic phosphorothioate. Isomer a shows the same K_m towards ribonuclease as uridine 2′ : 3′-cyclic phosphate, but V_{max} is lower by a factor of 5. (From Usher *et al.*, 1970.)

argument will not be repeated here; however compelling, arguments of this sort can never be wholly conclusive. However, since this proposal was made, Usher *et al.* (1970, 1972) in an elegant series of experiments have shown conclusively that the mechanism for both the transphosphorylation and the hydrolysis steps is indeed in-line. The key to this demonstration was the isolation by Eckstein (1968, 1970) of the two diastereoisomers of uridine 2′ : 3′-cyclic phosphorothioate (Fig. 8), which are both substrates for ribonuclease. In order to determine the geometry of the first step, Usher *et al.* (1972) took advantage of the fact that under appropriate conditions ribonuclease will catalyse the formation of a dinucleotide from a nucleoside and a nucleoside cyclic phosphate (Hepple *et al.*, 1955). Isomer a of uridine 2′ : 3′-cyclic phosphorothioate was incubated at −18° for 22 days with ribonuclease and a large excess of cytidine. The dinucleotide Up(s)C was isolated and reconverted to the cyclic phosphorothioate by treatment with

TABLE I

Hydrolysis of uridine-2′, 3′-cyclic phosphorothioate (isomer a) by ribonuclease A (from Usher et al., 1970)

Isomer isolated after ring closure (and purity)	Percentage incorporation: Expected if ring opening was adjacent (%)	Expected if ring opening was in-line (%)	Found (%)
a (about 90%)	86	14	6
b (about 75%)	28	72	70

The isomeric purity of the starting material was assumed to be 95% but may have been higher. Water used in the hydrolyses contained between 4 and 5% ^{18}O; analysis of the thiophosphate for ^{18}O incorporation was carried out by pyrolysis with mercuric chloride and cyanide, followed by mass spectrometric determination of the carbon dioxide produced on an isotope ratio machine.

potassium *tert*-butoxide in *tert*-butanol/formamide. The product was identified by ^{31}P n.m.r. as the a isomer, showing that the geometry of the ribonuclease reaction is the same as that of the base-catalysed reaction, which there is good reason to believe to be in-line.

To study the second step of the reaction, Usher *et al.* (1970) allowed the ribonuclease-catalysed hydrolysis of isomer a of uridine 2′ : 3′-cyclic phosphorothioate to proceed in ^{18}O-enriched water. The monoester produced was recyclised by diethyl phosphorochloridate in pyridine, and the ^{18}O

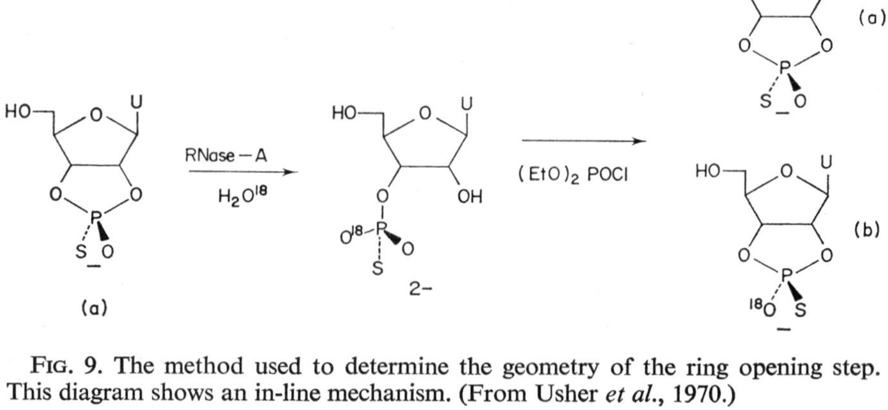

FIG. 9. The method used to determine the geometry of the ring opening step. This diagram shows an in-line mechanism. (From Usher *et al.*, 1970.)

content of the two separated isomers of the cyclic phosphorothioate determined. The incorporation figures expected (allowing for the purity of the isolated isomers) for the two mechanisms are compared with the observed incorporations in Table I, and the reaction scheme is shown in Fig. 9. Once again, the geometry of the ribonuclease-catalysed reaction is clearly the same as that of the chemical reaction, which is thought to be in-line.

These determinations of the geometry of the ribonuclease reaction clearly depend on the assumption that both the recyclisation by diethyl phosphorochloridate and the base-catalysed hydrolysis proceed by an in-line mechanism. Usher *et al.* (1970) demonstrated by an ^{18}O incorporation experiment that both these reactions have the *same* geometry, and the "preference rules" quoted above, which now have considerable experimental support (for reviews see Westheimer, 1968; Usher, 1969), indicate that this must be in-line (since the ring must span one apical and one equatorial position, and the relatively electropositive anionic groups $-O^-$ and $-S^{\delta-}$ will preferentially occupy equatorial positions).

The experiment of Eckstein *et al.* (1972), in which the absolute stereochemistry of uridine 3′-O-thiophosphate methyl ester produced by the action of ribonuclease on isomer a of uridine 2′ : 3′-cyclic phosphorothioate in 80% methanol was determined by X-ray diffraction, does not depend on a knowledge of the geometry of any other reaction. This experiment also shows that the geometry of the first step is in-line; it is of course subject to a possible ambiguity, in that methanol may bind in a different orientation from a nucleoside in the active site.

These experiments thus exclude any mechanism involving pseudorotation; that is, any mechanism in which the same group acts first as a general base toward the attacking group (2′O or water) and then as a general acid toward the leaving group (5′O or 2′O). The mechanism of Findlay *et al.* (1962) as modified by Roberts *et al.* (1969) is shown in Fig. 10 to illustrate the type of in-line mechanism which is consistent with the results of Usher *et al.* (1970, 1972). The most recent crystallographic results indicate that lysine 41 is probably not, in fact, in contact with the phosphate group in the dinucleotide or monoester complexes, though due to the movement of the phosphorus atom it might well be in the cyclic phosphate and perhaps in the pentacovalent intermediates (Richards and Wyckoff, 1971).

Although it is now possible to define with some accuracy the geometric course of the reaction catalysed by ribonuclease, we are still some considerable way from an understanding of the details of the catalytic mechanisms of this enzyme. Hammes and his colleagues (see Hammes, 1968; Hammes and Walz, 1969; Walz, 1971) in an extensive series of kinetic investigations have shown that there are conformational isomerisations of the free enzyme and of its complexes with CpC, cytidine 2′ : 3′-cyclic phosphate and cytidine

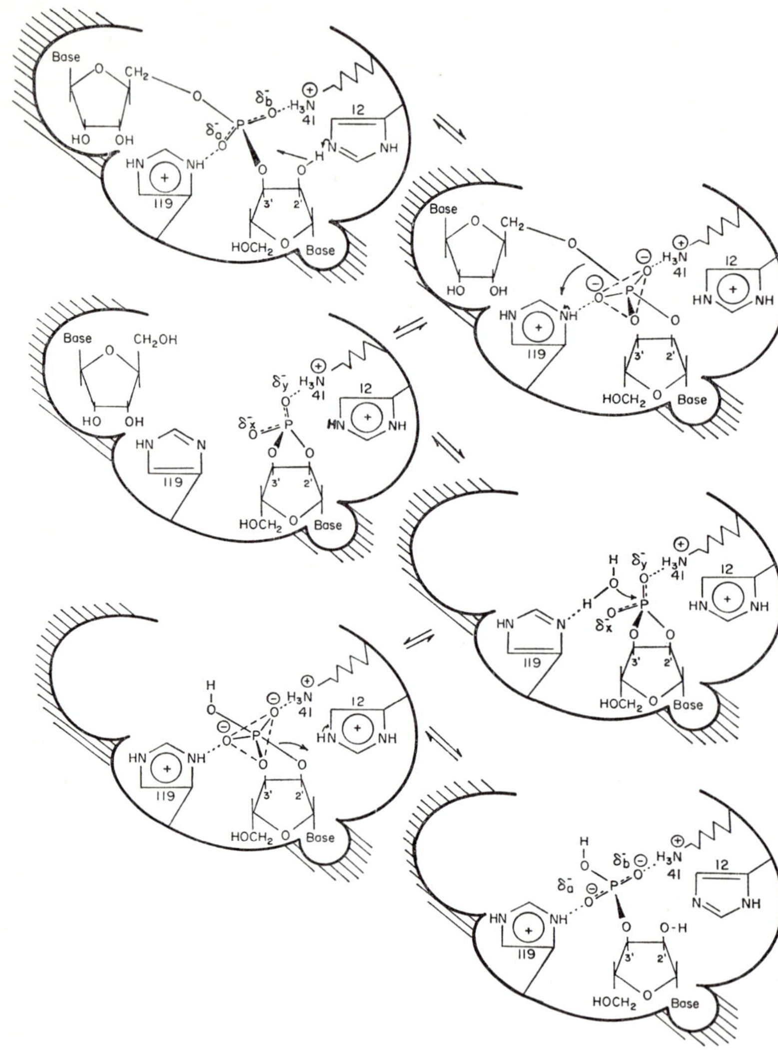

FIG. 10. Proposed mechanism of action of RNase. The relative positions of the substrate and the amino acid side-chains deduced from X-ray crystallography and n.m.r. spectroscopy are shown, as far as possible in two dimensions. The shaded areas represent binding sites for the two nucleoside bases. The notation δ^- is used to indicate the uncertainty in the negative charge on the phosphate oxygens. ($\delta_a^- \sim \delta_b^- \sim 1/2$) and ($0 < \delta_x^- < 1/2 < \delta_y - < 1$). In the last structure, the ionisation states shown are those which result from the reaction; optimal binding of the product results when the phosphate group is doubly ionised and histidine 12 is protonated. (From Roberts *et al.*, 1969.)

3′ phosphate. The minimal reaction scheme indicated by their experiments is shown in Fig. 11, where the various ionisation states of the enzyme and the nucleotides have been ignored for clarity. To this scheme one must add the pentacovalent intermediates discussed above, though no experimental information on their lifetime is available—Hammes (1968) has argued that it must be extremely short. A real understanding of the mechanism of ribonuclear will only be possible when the structure of each of these intermediates is known in crystallographic detail, and the pK values of the relevant groups have been determined (for example by n.m.r. : Meadows and Jardetzky, 1968).

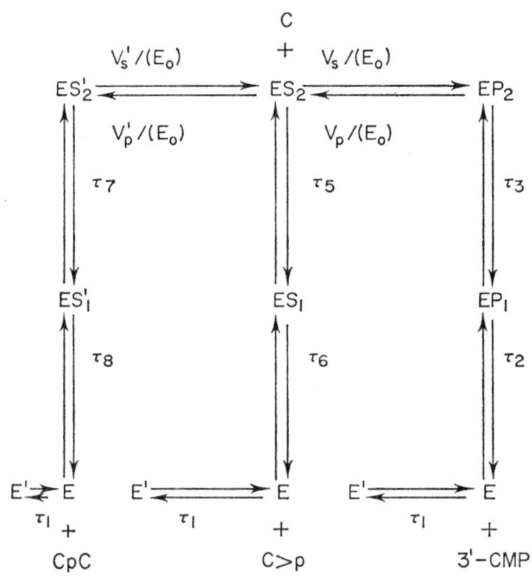

FIG. 11. Kinetic mechanism for the interaction of substrates and products with ribonuclease. Proton binding equilibria are not shown. The various isomerisation processes are discussed in Section V. Adapted from Hammes (1968), in view of the results of Hammes and Walz (1969).

Only the equilibrium complexes with the product, 3′CMP, and with a dinucleotide substrate analogue have been studied in detail at present. The recent report by Leinhard *et al.* (1971) that the uridine-oxovanadate (IV) complex is a powerful inhibitor of ribonuclease is particularly interesting in this regard. This complex was designed as a transition state analogue, since it contains a pentacovalent vanadium with either square pyramidal or trigonal bipyramidal geometry. The strong binding of the complex to ribonuclease ($K_i \simeq 7 \times 10^{-6}$ M) suggests that it is indeed a good analogue of the pentacovalent intermediate, and this opens up the prospect of a much better understanding of the geometry at this crucial stage of the reaction.

The present rather limited knowledge of the nature of the various conformational isomerisations of ribonuclease implicated by Hammes in the overall mechanism will be discussed in Section V.

IV. Correlation of the Information obtained by Other Techniques with the Model of Ribonuclease–Nucleotide Complex obtained from X-ray Crystallography

In this section we shall describe the conclusions drawn by protein chemists and spectroscopists concerning the mode of interaction between nucleotides and RNase A. Information on binding from chemical modification experiments, u.v., ORD/CD and n.m.r. spectroscopic experiments will be considered here. Information derived from these techniques concerning possible conformational changes will be discussed later. As in the section on specificity, experiments which seem to be important but are not at present readily explicable, will be considered briefly in Section VI. No attempt has been made to give an exhaustive coverage but most of the experiments providing detailed information on the binding site will be discussed.

A. CHEMICAL MODIFICATION

Long before the crystal structure was determined, elegant and pioneering chemical modification experiments clearly showed that histidines 12 and 119 and lysine 41 were intimately involved either in binding or catalysis and that the active site as a whole consisted of a cluster of these and possibly other positively charged residues. These and subsequent experiments have allowed the identification of residues at the active site, and some definition of their role in the function of the enzyme. As far as the histidines at the active site are concerned, the classical photo-oxidation experiments of Weil and Seibles (1955) showed clearly that enzymatic activity was abolished with the oxidation of histidine residues, without significant modification of the other amino acids present. Although this experiment did little to elucidate which particular histidine residues were associated with the activity loss, it focused attention upon them. Richards (1958) narrowed the search immediately by carrying out the photo-oxidation experiments separately on S-peptide and S-protein. Modification of either resulted in inactivation and since the S-peptide contains only His 12, this residue was quickly singled out as being necessary for activity. The S-protein portion contains the other three histidines so no assignment could be made at that time. However, after the importance of His 119 had been demonstrated by other modification reactions discussed below, it was subsequently shown to be the residue involved in the activity loss on the S-protein portion of the RNase molecule (Kenkare and Richards, 1966). Waku and Nakazawa (1965), noting that RNase was protected against the effect of γ-rays by substrate analogues (Ukita and Waku, 1964), carried out a similar experiment with the photo-oxidation reaction and showed that photo-oxidation of the histidines was inhibited by the presence of 2′(3′)-UMP or 2′(3′)-CMP but not by cytidine or adenosine nor by 2′(3′)-UMP at pH 8·0. They isolated an apparently homogenous material, which had full RNase activity, but on amino acid analysis, was found to lack two histidine residues. Based on the known importance of His 12 and 119 they concluded that the mononucleotide inhibitors had protected these residues and that His 48 and 105 were the histidines oxidised.

The first definitive experiments, which implicated His 119 in RNase activity, were the result of carboxymethylation with bromocetate (Barnard and Stein, 1959; Stein and Barnard, 1959) and with iodoacetate (Grundlach *et al.*, 1959; Crestfield *et al.*, 1963a, b). These experiments showed that a total loss of activity resulted with the introduction of one carboxymethyl group either on the N-1 of His 119 or on the N-3 of His 12. These reactions were found to be mutually exclusive and the major product was that carboxymethylated on His 119. Reaction at either site was reduced in the presence of competitive inhibitors and the inhibition of the carboxymethylation was directly related to their ability to inhibit enzymatic activity (Barnard and Stein, 1959; Ross *et al.*, 1962; Crestfield *et al.*, 1963a; Goren and Barnard, 1970). Comprehensive reviews of the large number of experiments carried out with this system are available (Barnard, 1969; Richards and Wyckoff, 1971) and only the conclusions concerning the active site and its geometry drawn from them will be mentioned here.

The very high rate of reaction of the active site histidine, its marked pH dependence and the abolition of this reactivity by disruption of the tertiary structure by denaturants or by oxidation or reduction of the disulphide bonds immediately demonstrated that both geometric and electrostatic factors were involved in this unique reactivity. The geometrical constraints were further exemplified by the experiments of Heinrikson *et al.* (1965) which demonstrated how the rates and ratios of carboxyalkylation at the two residues could be varied over a wide range by using haloacids of various sizes and stereochemistry. At the present time, no detailed correlation of Heinrikson's experiments with the crystal structure has been made.

Goren and Barnard (1970) demonstrated that the alkylation reaction in the 25–40°C range showed a tendency to level off, thus implying that reactivity was highly dependent on the exact geometry of the residues at the active site, which could be easily disrupted by increasing side chain mobility. Indeed, it was postulated that one histidine may orient the haloacetate ion for reaction with the other (Stark *et al.*, 1961; Lin *et al.*, 1968). The finding that His 12 is alkylated exclusively at N-3 and His 119 at N-1 also implied that geometrical constraints on reactivity are imposed by the tertiary structure of the enzyme.

In the X-ray model the N-3 group of His 12 is seen to be hydrogen bonded to the backbone carbonyl of Thr 45, while His 119 in position IV (Fig. 1) has its N-3 within hydrogen bonding distance of Asp 121, leaving N-1 free for reaction. The very slow reaction of His 12 with iodoacetamide and the pH dependence of the reaction suggested that the active site was probably in a region of positively charged amino acids but that a single carboxyl group may be nearby; the X-ray model shows Asp 121 to be close to His 119. Goren and Barnard (1970) were able to determine the pKa values of the two histidine residues by studying the pH dependence of the reaction and found them to be 4·8 and 6·2 for His 119 and His 12 respectively. Lamden *et al.* (1962) suggested that two groups with pKs of 4·85 and 5·55 were involved. These rather low pK values may contribute to the enhanced reactivity of these histidines compared to model compounds. The microscopic pK values for these residues determined by n.m.r. are discussed later. Further compelling evidence for the importance of both His 12 and His 119 in activity came from the dimerisation experiments of Crestfield and Fruchter (1967). These investigators found that the active site dimer of CM-12-RNase and CM-119-RNase, both of which are inactive alone, had 50% of the specific reactivity of native ribonuclease and this activity was completely destroyed on thermal dissociation of the dimer. They postulated that the dimers formed in such a way as to create one active site containing both CM-His 12 and CM-His 119, and a second unmodified site. A full description of the properties of these aggregates is given by Richards and Wyckoff (1971).

A further demonstration of the importance of histidine 12 for ribonuclease activity was the chemical synthesis of an S-peptide analogue which was identical to the native sequence of residues 1–14 except that histidine 12 was replaced by a β-(pyrazolyl-3)-L-alanine group (Finn and Hoffmann, 1967; Hoffmann *et al.*, 1971). When added to S-protein, this peptide failed to generate any enzymic activity. Earlier work had already shown that residues 15–20 were unnecessary for RNase S′ activity (Hoffmann *et al.*, 1966) and so the replacement of the imidazole ring at position 12 by the isosteric pyrazole ring was responsible for the complete loss of enzymatic activity. The binding of the modified peptide to S-protein was similar to that of the native peptide. Clearly the steric and/or electronic requirements for the catalytic function of His 12 are very stringent. Numerous additional synthetic analogues of S-peptide have been prepared which demonstrate the importance of various residues to either S-peptide binding, enzymatic activity or both and references for these can be found in either Richards and Wyckoff (1971) or in one of the most recent contributions by Scoffone's group (Borin *et al.*, 1972)

The possible importance of lysine residues in ribonuclease activity was first indicated by the guanidination experiments of Klee and Richards (1957). Reaction of RNase with O-methyl-isourea resulted in the rapid guanidination of about 9 lysine residues with only a very slow reaction of the tenth. Experiments with 100% and 95% guanidinated RNase indicated that probably only the last lysine to react was responsible for the loss of activity. This lysine was thought to be located at the active site of the enzyme since the binding of 3′-UMP markedly decreased the rate of inactivation. This work was subsequently reinvestigated by Glick and Barnard (1970) using a new reagent 1-guanido-3,5-dimethyl pyrazole. These authors showed that loss of activity did indeed result from the guanidination of the tenth lysine residue. The 9-guanidino-ribonuclease showed considerable enzyme activity and, like the native enzyme, was specifically inactivated at an unusually fast rate by fluorodinitrobenzene (FDNB). Since, as will be discussed below, FDNB is known to react rapidly with lysine 41, this lends strong support to the idea that the lysine residue which is guanidinated most slowly, and whose modification leads to loss of activity, is indeed lysine 41.

The specific and rapid inactivation of RNase by modification of lysine 41 with FDNB was demonstrated by Hirs (1962). Further experiments demonstrated that this reaction occurred some 70 times faster than expected for a lysine residue in a small peptide (Hirs *et al.*, 1965). The reaction was strongly inhibited by competitive inhibitors of ribonuclease activity in the order nucleotide > nucleosides > phosphate. Hirs estimated that the c-NH_2 group of lysine 41 might be *c.* 6 Å from the centre of the pyrimidine ring. This prediction would seem to find support in the X-ray structure. Murdock *et al.* (1966) studied this reaction in much more detail and from the pH dependence concluded that lysine 41 has a $pK_a = 8{\cdot}8$ which is much lower than that of a more normal lysine in a peptide or protein (pK_a *c.* 10·2). The simplest explanation for this is its location in an electropositive environment. The X-ray model shows arginine 39 directly above and lysine 7 across the cleft from lysine 41. Experiments with the bifunctional reagent 1,5 difluoro, 2,4-dinitrobenzene demonstrated that lysine 7 and lysine 41 are extremely close to one another in the tertiary structure as cross linking of these two residues was possible with this reagent (Marfey *et al.*, 1965).

Anfinsen's group also focused attention on the uniqueness of certain lysine residues (Anfinsen *et al.*, 1962). RNase was reacted with N-carboxy-DL-alanine anhydride which attaches chains of alanine residues to the ε-amino groups of the lysine. This material maintained virtually full enzymatic activity; however it was later shown that three of the lysines, numbers 7, 37 and 41, were resistant to reaction (Cook *et al.*, 1963). Even after exhaustive reaction lysine 41 remained completely unmodified. When the reaction buffer was changed from phosphate to bicarbonate however, lysine 41 reacted rapidly with a parallel loss in activity. The phosphate anion selectively protected lysine 41 from the reagent. This implies that lysine 41 is very close to if not actually part of the binding site for the dianionic phosphate ion. Wellner *et al.* (1963) confirmed these findings and carried out extensive activity measurements on various polyalanylated derivatives using both RNA and C>p as substrates as a function of pH and ionic strength; they concluded that polyalanylated ribonucleases had new properties of its own. A comparative study of basic, acidic, and neutral polypeptidyl ribonucleases was carried out by Frensdorff *et al.* (1967). Their studies showed that Lys 41 must be unmodified for the derivative to posesss enzymatic activity. In addition these authors focused attention on the possible existence of auxillary binding sites for polyanionic RNA which were absent on these modified enzymes. They observed that polyarginyl, polylysyl and polyornithyl RNase which possessed a "brush" of positively charged residues surrounding the native enzyme had lost virtually all of their ability to hydrolyse RNA but maintained full activity against C>p. They concluded this was probably due to the formation of abortive complexes of RNA with the surrounding positively charged "brush", precluding specific binding at the active site. Some evidence that Lys 7 may play a role as an auxiliary binding site in the native enzyme was presented by Rocchi *et al.* (1969) who synthesised a des-(Lys 7)S-peptide analogue and showed it to have decreased activity against RNA but not against C>p when combined with S-protein. Hoffman *et al.* (1971) disagreed with this conclusion. They also deleted Lys 7 in the S-peptide but replaced it with a norleucine residue and found this derivative to be able to generate activity in combination with S-protein. Their derivative was less effective in this regard than native S-peptide, however, so the question would seem to remain open. It was found that RNase S′ formed with 1α, 1ϵ, 7ϵ triacetamido-S-peptide had only 40% activity against RNA whereas 1α, 1ϵ, 7ϵ triguanido-S-peptide maintained full activity, implying that positive charges in these positions are involved in RNA binding. Observations by Eaker (1961) on des(lys 1)-RNase, by Finn and Hoffmann (1965) with S-peptide 2–13, and by Moroder *et al.* (1969) with des(lys 1) [Orn 10]-S-peptide would seem to rule out the involvement of Lys 1 in such interactions.

Recent synthetic work by Hoffman *et al.* (1971) and Marchiori *et al.* (1972) has provided evidence for an electrostatic interaction between Glu 2 and Arg 10 in the S-peptide portion of RNase S′, which may be important in maintaining His 12 in the proper orientation for maximum activity. By demonstrating that the k_s values were more dependent upon this Glu 2–Arg 10 interaction than were the K_m values for substrate or the K_i values for 3′-CMP, Marchiori concluded that the Glu 2–Arg 10 interaction operates primarily to maintain catalytic function rather than binding ability. Recently Finn *et al.* (1972) have provided independent evidence for this interaction by n.m.r. spectroscopy.

Similarly, possible influences on this orientation of His 119 have been studied by proteolysis experiments. Thus, while removal of Val 124 on S-protein by carboxypeptidase had little or no effect on the activity against RNA of the subsequently reconstituted RNase S′ (Potts *et al.*, 1964; Richards and Wyckoff 1971), removal of Ser 123 by more extensive digestion with carboxypeptidase results in a decrease in activity to about half that of normal RNase S′, and removal of residues through Phe 120 resulted in a loss of all activity even though His 119 was still present. Anfinsen (1956), using a limited pepsin digestion of RNase at low pH, was able to isolate a homogeneous protein which he showed to lack the C-terminal residues 121–124. This material has subsequently been called des(121–124)RNase (Anfinsen, 1956; Richards and Wyckoff, 1971). Although it was originally thought that the derivate was inactive, a detailed analysis by Lin (1970) on carefully purified material showed it to have a K_m of 0·5 mM, two-fold greater than the native enzyme, and a turnover number of 2×10^{-3} sec^{-1} (only 0·5% of the value for the native enzyme). In addition the K_i value for 2′-CMP was found to be some thirteen-fold greater in des(121–124)RNase. Alkylation of His 12 by iodoacetate was accelerated in this derivate but the reactivity of His 119 was severely diminished (Lin *et al.*, 1968). From these data the authors concluded that Ser 123 can have only a minor role in the binding properties of RNase and that the low enzymatic activity and decreased susceptibility to iodoacetate may be better explained by a change in the pK_a of His 119, its orientation in the derivative, or both. Examination of the X-ray model suggests that these effects are likely to be due to the removal of Asp 121 which, like the removal of Glu 2–Arg 10 interaction mentioned above, seems primarily to influence the catalytic efficiency of the active site rather than its binding properties. What little enzymatic activity or ability to bind nucleotides remains in des(121–124) RNase is completely eliminated upon removal of Phe 120 with carboxypeptidase. As this derivative has now lost much of the stability of the des(121–124) RNase molecule, detailed analysis of this result is difficult but the simplest interpretation seems to be that Phe 120 must be inserted into the hydrophobic region below the active site to maintain His 119 in a proper position for catalysis. This region may in fact be important for maintaining the overall structure of the molecule.

In conclusion it can be said that virtually all the predictions about the environment of individual residues made from these modification experiments are confirmed (and of course refined) by the crystal structure. However, it should also be clear that much of the information obtained from these experiments could not be derived, *a priori*, by simple examination of the X-ray structure. Glu 2 is sufficiently close to interact with Arg 10 in the X-ray model but one could not predict that this interaction would be important for maintaining optimal active site geometry. Similarly although His 12, His 119 and Lys 41 are all together at the active site in the X-ray model, one could not have predicted the unique reactivities observed for these residues.

The data from protein modification experiments clearly supplement the X-ray information and will continue to be important to a detailed knowledge of the factors influencing the interaction of the protein with nucleotides.

B. ULTRAVIOLET ABSORPTION SPECTROSCOPY

In this section we will not consider the large amount of data concerning the u.v. spectrum of RNase itself and its changes on denaturation and renaturation of the enzyme, but only the information concerning ligand binding which has been derived from u.v. spectroscopy.

Mathias *et al.* (1960), Findlay *et al.* (1961) and Ross *et al.* (1962) observed that the u.v. spectrum of a mixture of RNase with 2′-CMP (and to a lesser extent with 3′-CMP), in the 250–280 nm region, was less than the simple computed addition spectrum of the two substances alone. The λ_{max} was also shifted to longer wavelengths. These findings are in agreement with a similar set of results obtained by Hummel *et al.* (1960). The former authors, however, suggested that CM-His 119 ribonuclease could also bind inhibitors since a non-additive u.v. spectrum was generated upon mixing with 2′-CMP or C>p. The presence or absence of inhibitor binding to carboxymethyl derivatives or ribonuclease has been a matter of some controversy and will be discussed later. Hummel *et al.* (1961) subsequently reported a detailed study of the ribonuclease–nucleotide interaction by u.v. spectroscopy. Their difference spectra clearly showed a marked decrease in absorbance between 240–270 nm and in addition showed more subtle variations at higher wavelengths which depended on the particular nucleotide under consideration. The order of decreasing magnitude of the difference spectra were 2′-CMP > 2′deoxy3′-CMP > 3′-CMP > 2′ and 3′-UMP > 2′-AMP > 3′-AMP. Since neither cytidine nor the benzyl or methyl esters of 2′-CMP generated difference spectra these authors concluded that a phosphomonoester group at the 2′ or 3′ position of the nucleotide was necessary. However, the fact that neither 2′ nor 3′-GMP, both of which are inhibitors of RNase activity, gave appreciable difference spectra, suggested that this was not the only requirement. The binding stoichiometry for 2′-CMP was shown to be 1 : 1; maximum amplitudes of the difference spectra were observed at pH 5·5 and a group with a $pK_a = 6{\cdot}5–6{\cdot}8$ was implicated in the binding process. It was also observed that the wavelength of maximal spectral change was related to the λ_{max} of the particular nucleotide in question, suggesting that the difference spectra were primarily due to a red shift of the absorption of the nucleotide with little or no change in the absorption of the aromatic groups on the protein. Hummel *et al.* (1961) also pointed out, however, that the slight inflection in the difference spectra at 280·5 nm and 286 nm might reflect small perturbations in one or more of the tyrosine residues of the protein.

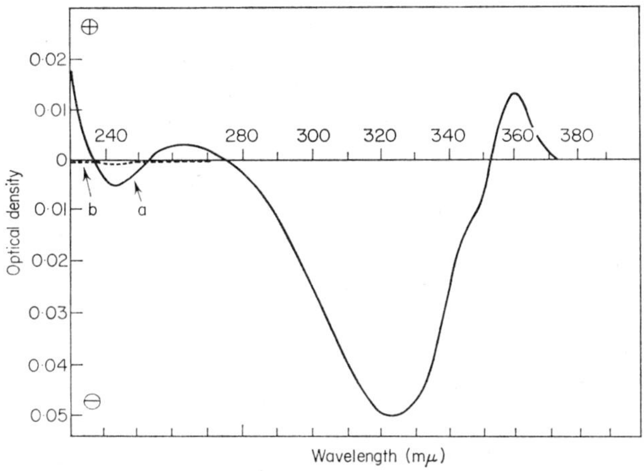

FIG. 12. Difference spectra observed upon mixing 4-thiouridine-2′(3′) phosphate and 5,6-dihydrouridine-2′(3′) phosphate with RNase A. Curve a: Buffer acetate ($\mu = 0{\cdot}01$, pH 5·6). Final concentrations of both enzyme and 4-thiouridine-2′(3′) phosphate were 0·05 mM. Curve b: Buffer acetate ($\mu = 0{\cdot}02$, pH 5·5). Final concentrations of the enzyme and 5,6-dihydrouridine-2′(3′) phosphate were 0·06 mM and 10 mM respectively. (From Irie and Sawada, 1967.)

Irie and Sawada (1967) presented evidence to indicate that tyrosine residues are not, in fact, involved. Using 5,6-dihydrouridine 2′(3′)phosphate, which does not absorb above 240 nm, they failed to observe a difference spectrum when sufficient nucleotide had been added to form a 1 : 1 complex (Fig. 12). In addition, with 4-thiouridine-2′(3′) monophosphate, which has a λ_{max} at 331 nm, they observed a peak in the difference spectrum at 350 nm rather than at 280 nm (Fig. 12) suggesting that the inflection at 280 nm observed with the uridine and cytidine nucleotides is attributable to changes in nucleotide absorption. Sawada and Ishi (1968) showed that the interactions of 2′(3′)UMP and of 2′(3′)4-thio-UMP with ribonuclease are closely similar, so that the latter's use in experiments such as these seems entirely justifiable. Further evidence that at least the exposed tyrosine residues of ribonuclease located at positions 73, 76 and 115 are not involved, was furnished by the fact that no difference spectrum was generated in the region of nitrotyrosine absorption (360 nm) when 2′(3′)UMP was bound to a derivative of ribonuclease in which three tyrosine residues had been nitrated (Irie and Sawada, 1967).

More recently attempts have been made by Irie (1968), Deavin *et al.* (1968) and Anderson *et al.* (1968) to obtain more detailed information on the environment of the bound nucleotides from an analysis of these difference spectra. Irie postulated that the observed red shift and hypochromicity in the nucleotide spectrum could be accounted for by a change in the refractive index of the environment of the nucleotides and a stacking interaction between the base and an aromatic side chain at the active site. These conclusions were reached by studying the effects of variations in the solvent refractive index and the addition of aromatic compounds on the u.v. spectra of the nucleotides. Artificially shifting the spectrum of 2′-CMP or -UMP by 2 nm with a 20% hypochromism yielded a difference spectrum qualitatively similar to the difference spectrum observed for the complex with the enzyme. These conclusions are in reasonable agreement with the information on the structure of the complex from X-ray crystallography. Although in the X-ray model the pyrimidine base in site B_1 does not have its plane parallel to that of Phe 120, a weak stacking interaction might be possible. The base is also seen to be flanked on three sides by protein side chains and thus the postulated change in refractive index seems quite likely. Deaven *et al.* (1968) carried out similar experiments in an attempt to mimic, by a change in solvent and pH, the effects of binding to the enzyme on the u.v. spectrum of nucleotides; they concentrated on changes in charge state of the pyrimidine ring rather than stacking interactions as possible explanations of the difference spectra. As can be seen in Fig. 13, the magnitude of the difference spectra for the nucleotides of uridine are in the order of 5′-UMP > 5′-UDP > 5′-UTP, which parallels

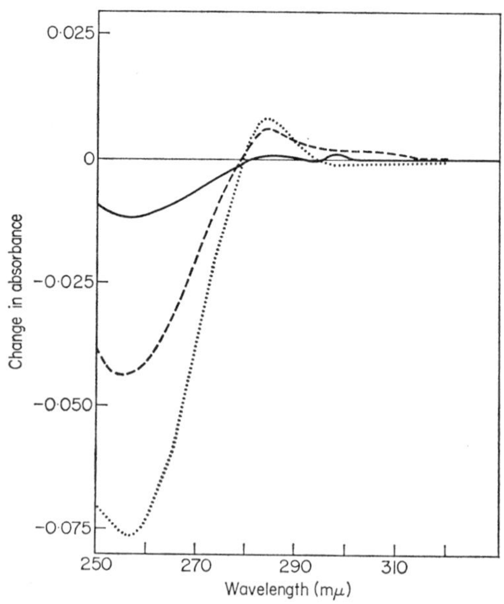

FIG. 13. Difference spectra of the interaction of ribonuclease and uridine 5′-phosphate (——), uridine 5′-diphosphate (– –), and uridine 5′-triphosphate (. . . .). Temp. 25°C, pH 5·20, I=0·02. Final concentration of protein and nucleotides after mixing were 50 μM. Cell 1 cm total path length. (From Deavin *et al.*, 1968.)

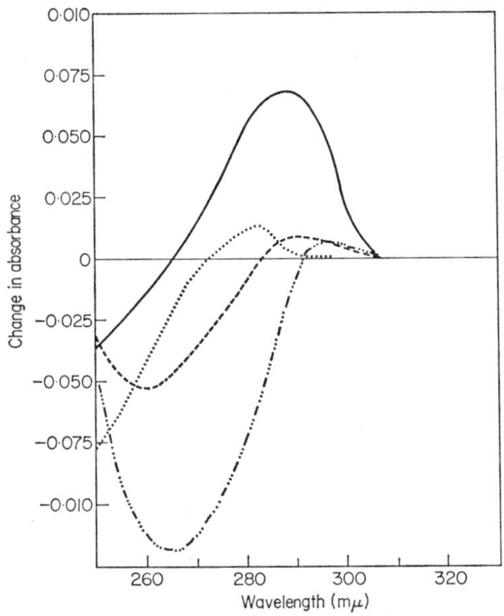

FIG. 14. Difference spectra resulting from a change in the pH of solutions of nucleotide. —···—, uridine-5′ phosphate final concentration after mixing 56 μM: pH changed from 6·7 to 10·5. —— cytidine 5′ phosphate final concentration 10 μM: pH changed from 6·4 to 1·82. ...... uridine, final concentration after mixing 50 μM: aqueous 0·05 M Tris buffer, pH 7·4 to 50% dioxan (v/v). – – – – – – uridine final concentration after mixing 50 μM: aqueous to 0·05 M borate buffer, pH 9·15, with 20% dioxan added. (From Deavin *et al.*, 1968.)

their binding constants. A comparison of the spectra in the figure with those in Fig. 14 reveals that a difference spectrum qualitatively similar to that observed on binding to ribonuclease can be generated by decreasing the pH and adding dioxane (50% v/v) to the free nucleotide. This observation was interpreted as evidence for a bond between either the 3 NH group or the 4-OH group of the uridine base (depending on its tautomeric form) and a carboxylate group on the enzyme. The aliphatic hydroxyl groups of Thr 45 and Ser 123 are thought to be the interaction sites for the 3 and 4 positions of pyrimidine bases in site B_1 (see Section II). However, in the free enzyme, the hydroxyl group of Thr 45 may be hydrogen bonded to the carboxylate of Asp 83. Such a system might explain the u.v. difference spectra but at the moment is pure speculation. In contrast with these results with uridine nucleotides, simple changes in the environment of free cytidine nucleotides could not mimic the perturbation of the u.v. spectra of these nucleotides on binding to ribonuclease. Ribonuclease-5′-GMP difference spectra could be mimicked by acidification of solutions of the free nucleotide. This was interpreted as evidence for the interaction of N-7 on the purine ring with an acidic group on the enzyme. Richards and Wyckoff (1971) pointed out that this effect could also be explained by an increase in the pK of N-7 brought about by the proximity of carboxylate anion(s), with Asp 111 and possibly Asp 121 being logical choices, since purine nucleotides bind primarily to site B_2R_2. Adenine nucleotides were also found to bind to the enzyme and to generate difference spectra. Acidification of the free adenine nucleotides generated a difference spectrum which apparently produced a splitting of the peak. However, in the bound nucleotide difference spectrum the intensity of the two components of the split peak are reversed. This was interpreted as possibly reflecting an equilibrium between a protonation on N-1 or on N-7 of the adenine ring, with one form being stabilised by the enzyme. Since little is known as yet about the specific interactions between the adenine ring and site B_2 on RNase no direct correlation between these u.v. data and the X-ray model can be made at this time.

Anderson *et al.* (1968) also studied the binding of pyrimidine mononucleotides to ribonuclease by u.v. difference spectroscopy and made a number of interesting observations. From the pH dependence of binding, they concluded that 3′-CMP binding was influenced by two ionizable groups ($pK_a = 5{\cdot}0$ and 6·5) whereas 2′ CMP binding was affected by three such groups (pK_a 5, 4·5 and 6·5). Although they observed general similarities in the u.v. difference spectra generated by the binding of a variety of pyrimidine mononucleotides, variations in the magnitude of the difference spectra led them to conclude that the base group of 2′ nucleotides interacts with the enzyme in a different manner from that of 3′ nucleotides. Crystallographic and n.m.r. evidence supporting this possibility is discussed later. Anderson *et al.* (1968) found no evidence that protonation of N-3 on the cytidine ring had any effect on the interaction of the nucleotides with the enzyme. Indeed the pK_a of N-3 in the bound state was identical to that in the free state. It is not easy to reconcile this result with the specific interaction between N-3 and the hydroxyl group of Thr 45 postulated from the X-ray model. Although this hydroxyl group can, of course, be a hydrogen bond donor or acceptor and could conceivably hydrogen bond to a protonated cytidine as it is believed to do to normal uridine, one would expect some difference in the strength of the interaction, and thus some change in the pK_a of N-3.

C. NUCLEAR MAGNETIC RESONANCE SPECTROSCOPY

Nuclear magnetic resonance spectroscopy has only recently been applied to the study of biological macromolecules and their interaction with ligands. Sensitivity limitations in the method of signal detection and the almost overwhelming complexity of the signals detected were the prime factors in the delay. With the advent of Fourier transform techniques, higher magnetic field strengths and methods for selective isotopic substitution, the full potential of this technique in modern biochemistry is beginning to be realised.

Although RNase was the first protein to be studied by proton n.m.r. spectroscopy (Saunders *et al.*, 1957), the first attempt to obtain detailed information from the spectrum was made by Bradbury and Scheraga (1966). With the knowledge that the C-2 proton resonances of histidine residues in RNase appeared at lower field than the main aromatic signals and that the chemical shift of the C-2 proton signal of histidine itself was dependent on the ionization state of the histidine ring, they collected the first n.m.r. results on the titration behaviour of the histidine residues in this enzyme. This work was extended in a detailed study by Meadows *et al.* (1967, 1968) which led to the assignment of the signals to specific histidine residues in the ribonuclease sequence and to the determination of the microscopic pK_a values of these residues. Assignment of the signals was based on (a) the carboxymethylation reactions of His 12 or His 119 discussed earlier, (b) selective deuteration of His 12 in RNase S by the incubation of S-peptide in D_2O, pH 7 at 40°C followed by recombination with S-protein to observe which peak was "missing", and (c) correlation of an anomalous chemical shift of one histidine C2-H resonance with the finding from the crystallographic studies that only histidine 48 is buried and relatively inaccessible to solvent. From low to high field the 4 histidine signals were assigned to residues 105, 12, 119 and 48 with pK_a values of 6·7, 6·2, 5·8 and 6·4 respectively (in 0·2 M sodium acetate, 32°C). The C-4 proton of His 105 was also observed and assigned to this residue by virtue of its pK_a being 6·7. Rüterjans and Witzel (1969) found that the effect of ionic strength on these pK_a values was different for each histidine; they interpreted this as evidence for varying degrees of electropositivity in the environment of each histidine. The assignments depended on assumptions concerning shift and pK_a changes for certain histidines which take place on conversion of ribonuclease to RNase S. Bradbury's group have recently confirmed the assignment given above by independent methods (Bradbury and Chapman, 1972; King and Bradbury, 1971).

After the assignment of the histidine resonances, Meadows and Jardetzky (1968) studied the effects of adding the product inhibitor 3′-CMP. The binding of 3′-CMP to ribonuclease causes selective downfield shifts in the resonances of His 12 and 119. His 105 is unaffected, whereas His 48 at pH 5·5 shows a 7 Hz shift downfield and displays variable linewidths at intermediate concentrations of 3′-CMP. As the concentration of 3′-CMP is raised the resonance of His 119 shifts a total of 60 Hz at pH 5·5 and 82 Hz at pH 7·0, with an increase in linewidth from 4–5 Hz to 8–9 Hz. Similarly His 12 shifts 12 Hz at pH 5·5 and 69 Hz at pH 7·0, but with constant linewidth. The pK_a value of His 119 in the complex is raised from 5·8 to 7·4 and that for His 12 from 6·2 to at least 8·0. The titration curves for the C-2 protons of His 119

TABLE II

(*Summary of chemical shift and pK changes on binding of inhibitors to RNase*† (*from Meadows* et al., 1969)

Inhibitor	RNase absorptions							Aromatic shift	Inhibitor absorptions at pH 5·5		
	His 12		His 48	His 105		His 119					
	pK	Δδ‡	Δδ‡	pK	Δδ‡	pK	Δδ‡		C(5)—H	C(6)—H	C(1′)—H
3′—CMP	8·0	+10	−7	6·7	0	7·4	−20	+30	−6	−22	−3
2′-CMP	8·0	+8	−10	6·7	0	>8·0	−25	Yes§	−4	≃ −20	−12
5′-CMP	8·0	+10	−10	6·7	0	<7·0	0	+30	−5	−17	−3
Cytidine	6·2	0	−10	6·7	0	6·4	0	Yes	−5	−18	−3
5-methyl-2′-deoxycytidine	6·2	0	−5	6·7	0	6·2	0	Yes	−5	−18	−3
Phosphate	6·9	0	−6	6·7	0	6·6	+10	No			
Sulphate	6·6	0	0	6·7	0	6·2	+10	No			

† Downfield chemical shift changes are given as negative numbers, and are in Hz. In all cases except cytidine and 5′methyl-2′-deoxycytidine inhibitor concentrations were sufficient to saturate the enzyme between pH 5 and pH 7.

‡ Δδ = change in chemical shift of the C(2)-H peak of the fully protonated histidine.

§ Due to broadening of the peak, this shift could not always be followed over the whole concentration range.

and His 12 in RNase saturated with 3′-CMP indicate that at low pH the shift of 119 is 20 Hz further *downfield* and His 12 is 7–10 Hz further *upfield* than the normal chemical shift values for these residues in their protonated form in the absence of inhibitor. Observations on the main aromatic envelope, which contains the resonances of the six tryosine and three phenylalanine residues of ribonuclease, indicate an upfield shift of a peak with the area of about 5 protons on increasing the concentration of 3′-CMP. This area could correspond to a single phenylalanine residue or the upfield doublet of at least two tyrosine residues. These data were interpreted as follows. The downfield shifts of the resonances of His 12 and 119 on inhibitor binding are primarily due to the protonation shift arising from the increase in the pK_a values of these two residues in the inhibitor complex. The additional 20 Hz downfield shift of His 119 is thought to be due to an interaction with the phosphate dianion of the inhibitor. This downfield shift of His 119 is similar to that observed for the interaction of imidazole and phosphate in DMSO (Cohen, 1968). The precise cause of the shift is not known; one possibility is the anisotropic deshielding effect of phosphorus–oxygen double-bond. His 105 was not affected by the inhibitor and thus appears not to be involved in the binding process. The shift and linewidth behaviour of His 48 was consistent with 3′-CMP binding affecting the equilibrium position of a pH dependent isomerisation of the free enzyme (see Section V).

This initial investigation was followed by a more complete study of the interaction of a number of inhibitors with the enzyme (Meadows *et al.*, 1969). A summary of the results of these experiments can be found in Table II. This table summarises (a) the changes in pK_a for each histidine observed on binding of various inhibitors, (b) the changes in chemical shift in the fully protonated state, (c) the absence or presence of the perturbation in the aromatic region, and (d) the changes observed in the resonances of the ligand at low ligand/enzyme ratios. A summary of the conclusions drawn from these inhibition interactions will be presented here, considering primarily the differences in binding of the various inhibitors. For a full description of the results the original paper should be consulted (Meadows *et al.*, 1969).

From the pH dependence of the chemical shifts of His 12 and 119 in the presence of various ligands it was concluded that both of these residues were positively charged in the complexes with 2′-CMP and 3′-CMP but that His 119 could deprotonate without causing dissociation of the complex with 5′-CMP. In the simple phosphate or sulphate complexes, both His 12 and 119 appeared to be able to deprotonate in the complex. Although Meadows and Jardetzky (1968) and Meadows *et al.* (1969) provided only indirect evidence that the dianionic form of the mononucleotides is the species which binds, subsequent experiments by Rüterjans *et al.* (1971) lend strong support to this hypothesis. These latter authors studied the pH dependence of the ^{31}P resonances of various nucleotides bound to RNase A and showed that for 2′-CMP, 3′-CMP and 2′-CMP N-oxide the chemical shift of these resonances, in the pH range where complex formation was maximal, corresponded to those of the dianionic state of the free nucleotide (Fig. 15). Only a small shift in the pK of 5′-CMP was observed, which supports the conclusion that His 119 does not interact strongly with the phosphate of 5′-CMP (as was discussed above). Meadows *et al.* (1969) proposed that a direct interaction between the phosphate dianion and His 119 made a significant contribution to the binding energy between mononucleotides and RNase since the binding strength of the

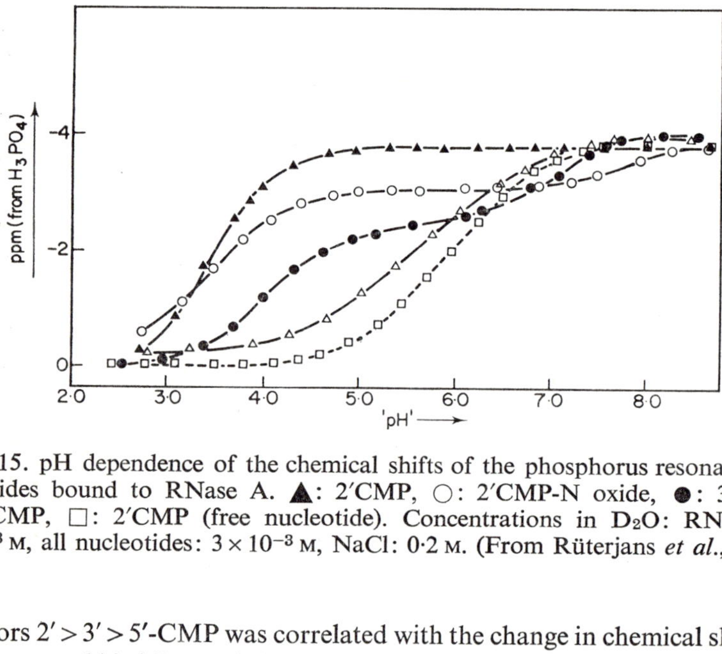

FIG. 15. pH dependence of the chemical shifts of the phosphorus resonances of nucleotides bound to RNase A. ▲: 2′CMP, ○: 2′CMP-N oxide, ●: 3′CMP, △: 5′CMP, □: 2′CMP (free nucleotide). Concentrations in D_2O: RNase A: 3×10^{-3} M, all nucleotides: 3×10^{-3} M, NaCl: 0·2 M. (From Rüterjans *et al.*, 1971).

inhibitors 2′ > 3′ > 5′-CMP was correlated with the change in chemical shifts of this protonated histidine and the apparent pK_a value of His 119 in the presence of the inhibitor (Table II). There was no evidence from the n.m.r. experiments to suggest that His 12 interacts directly with the mononucleotide dianion. In fact the observation that the apparent pK_a of His 12 was unaffected by the different orientations of the phosphate group in the 2′, 3′ or 5′-CMP complexes, and that the change in chemical shift of the fully protonated His 12 was in the opposite direction of that seen for His 119 or for imidazole hydrogen bonded to a phosphate group in model compounds (Cohen, 1968) supported the opposite conclusion. With the aid of the X-ray model, Meadows *et al.* (1969) were able to rationalise the shifts observed for the cytidine ring protons and for the aromatic peak on the enzyme. The n.m.r. shifts observed were consistent with the proximity of the base and phenylalanine 120 as illustrated earlier in Section II.

In addition to the determination of the microscopic pK_a values and their changes on inhibitor binding, a second interesting observation was made by Jardetzky's group, concerning the conformation about the glycosidic bond of the nucleotides when bound to the enzyme. With the aid of the X-ray model Meadows *et al.* (1969) constructed a series of schematic drawings depicting the geometry of the bound nucleotides in the active site of RNase which best fit the X-ray and n.m.r. data. In the 3′-CMP complex, the cytosine ring was shown in the *anti*-conformation with respect to the ribose ring. In the 2′-CMP complex, however, a rotation about the glycosidic bond was necessary in order to maximise the interaction between the 2′-phosphate group and His 119, as indicated by the n.m.r. shifts, if the hydrogen bonding scheme for the base in site B_1 was to be maintained. This rotation places 2′-CMP in a conformation close to *syn* which is distinctly different from the conformation it is believed to have in solution (references cited in Meadows *et al.*, 1969). Further discussion of this effect is reserved for section V.

In the X-ray model discussed earlier, a potential interaction between the 2′-OH group of the sugar ring and His 12 was indicated as being geometrically possible. N.m.r. evidence for the absence of any such interaction with mononucleotides has recently been presented (Rüterjans *et al.*, 1971). These authors examined the titration behaviour of the histidine resonances of RNase in the presence of 3′-CMP and 2′deoxy-3′-CMP and observed that the titration curve for His 12 in the RNase A-3′-dCMP complex is identical to that in the RNase A-3′-CMP complex. They interpreted this as evidence against the proposed mechanisms of RNase action (discussed previously) which involve a His 12-ribose 2′-OH interaction. In this regard it should be emphasised that the ionisation state of the histidine residues at the active site of RNase in the presence of dianionic product inhibitors gives absolutely no information concerning the ionisation states of these residues in the presence of monoanionic 3′-5′ phosphodiester or cyclic phosphate substrates. From Meadows' experiments it should be clear that the precise orientation of the phosphate dianion group has a dramatic effect on the pK_a value of His 119. It is not difficult to imagine then that the precise orientation of the phosphate monoanion group in the enzyme-substrate complex could have a similar controlling effect on the pK_a values of His 119, His 12 or both. Thus extrapolation from the pK_a values observed in mononucleotide complexes to the situation in substrate complexes must be considered unreliable.

A recent n.m.r. study obtained kinetic rather than structural information concerning inhibitor binding to ribonuclease. Taking advantage of the large chemical shift range for ^{31}P resonances which extends the lower limit of the time scale over which exchange phenomena can be measured, Lee and Chan

(1971) found that the apparent dissociation rate constant of the 3′-UMP-RNase A complex at 32°C, pH 7·5 (0·1 M Tris, 0·05 M NaCl) was $3{\cdot}2 \times 10^3\ sec^{-1}$ with an activation energy of 3·5 kcal/mole. The value for this rate was determined by temperature jump studies was $1{\cdot}64 \times 10^4\ sec^{-1}$ at pH 7·0, 25°C and 0·2 M KCl (Hammes and Walz, 1969).

D. OPTICAL ROTATORY DISPERSION AND CIRCULAR DICHROISM

As in the other sections of this review we will not consider here the detailed interpretation of the ORD/CD spectra for the native enzyme but rather discuss how changes in these spectra have been used to monitor conformational changes in either the enzyme or ligand. This section will serve as an introduction to the following one concerning evidence for ligand induced conformational changes obtained by other techniques.

The main features of the optical rotatory dispersion curve between 220–300 nm for ribonuclease A are the Cotton effect centred at 278 nm and the large trough centred at 227 nm. Cathou *et al.* (1965) observed that in the presence of 2′-CMP and 3′-CMP the Cotton effect at 278 nm, which is probably associated with the tyrosine residues (Hooker and Tanford, 1964), disappeared. Pyrophosphate ion, which is also an inhibitor of ribonuclease, does not alter the 278 nm Cotton effect. The absence of the 278 nm Cotton effect in the presence of 2′ or 3′-CMP was interpreted as evidence for an

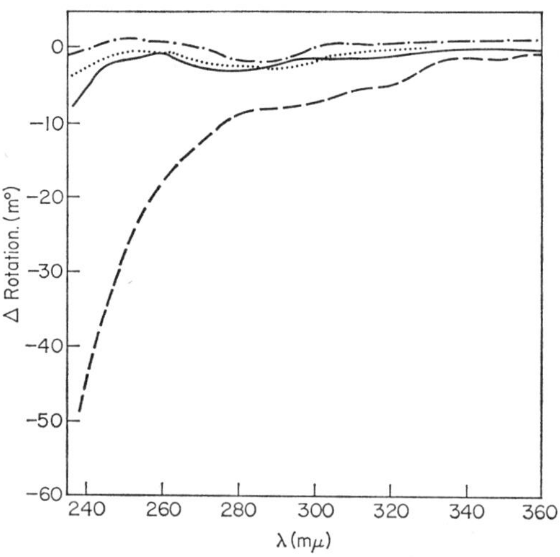

FIG. 16. Optical rotatory dispersion spectra of nucleotides ($1{\cdot}0 \times 10^{-4}$ M) and ribonuclease ($0{\cdot}5 \times 10^{-4}$ M) in a 1-cm cell. 0·02 M sodium acetate buffer pH 5·2, 25°C. ---- 3′-CMP; —— ATP; -·-·- 2′-deoxythymidine 3′ phosphate benzyl ester; ..... pyrophosphate. (From Deavin *et al.*, 1966.)

alteration in the environment of the tyrosines, with or without a significant change in the overall conformation of the protein. The ORD curves for ribonuclease were found to be pH independent over the range pH 5–8, whereas relaxation kinetic data suggested that the enzyme undergoes a pH-dependent isomerisation at neutral pH (French and Hammes, 1965). Cathou *et al.* (1965) concluded from this that the isomerisation observed by temperature jump kinetics cannot involve a massive conformational change.

Deavin *et al.* (1966) had been intrigued by the marked variation in transphosphorylation rates observed for 3′-5′-dinucleoside phosphates which are a function of the nature of the base on the 5′ end of the dinucleotide (Witzel and Barnard, 1962). (These rate differences will be discussed in section VI.) In order to gain some insight into the nature of the possible binding sites for dinucleoside phosphates, these authors studied the ORD difference spectra for various RNase–ligand complexes. The spectra for a few of these ligands are shown in Fig. 16. These spectra represent the calculated difference between the enzyme–ligand complex and the computed sum of the separate components. Negligible effects were observed in the complexes with ATP, deoxythymidine 3′-phosphate benzyl ester and pyrophosphate, but in contrast, rather large differences were observed in the ribonuclease–3′-CMP

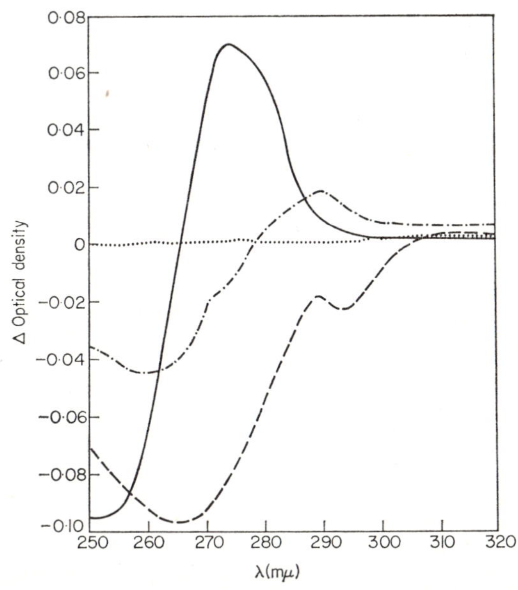

FIG. 17. Ultraviolet absorption difference spectra measured in split compartment cells (total path length 1 cm) in 0·02 M sodium acetate buffer pH 5·2, 25°C. All concentrations $0{\cdot}5 \times 10^{-4}$ M. —— 3′-CMP; —— ATP; –·–·– 2′ deoxythymidine 3′ phosphate benzyl ester; ..... pyrophosphate. (From Deavin *et al.*, 1966.)

complex especially below 280 nm. Similar effects to those seen for 3′-CMP were observed with 2′-CMP, 5′-CMP, 2′-UMP and 5′-UMP–RNase complexes. All of the ligands tested were able to combine with the enzyme as evidenced by their ability to generate u.v. difference spectra on mixing with enzyme (Fig. 17). Pyrophosphate has no nucleoside portion and did not generate a difference spectrum. This supports the evidence cited in Section IVB, that perturbations in the base are primarily responsible for the u.v. difference spectra observed in RNase–nucleotide complexes. Deavin *et al.* (1966) suggested that the unifying feature of those nucleotides capable of generating a difference ORD spectrum, which they equated with ligand-induced conformational changes on the enzyme, was that they possessed a 2′OH group on the ribose ring. Although this study does not prove that an interaction between the 2′OH group on the ribose ring and a group on the enzyme is responsible for the conformational change, it does support this proposal. The results concerning the importance of this interaction should be compared to the results of Rüterjans *et al.* (1971), discussed in the n.m.r. section, which indicate that the presence or absence of a 2′OH group has

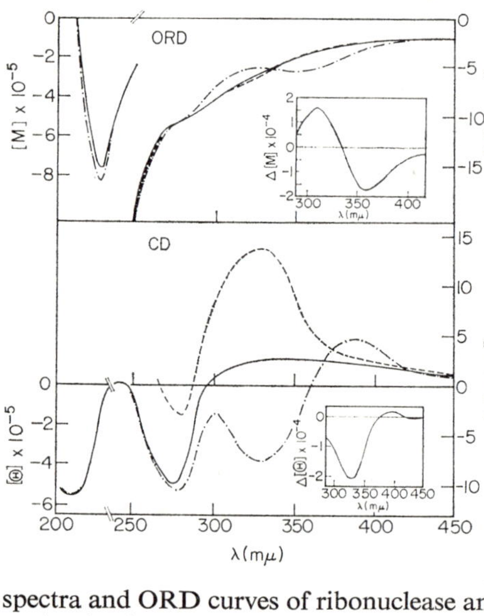

FIG. 18. The CD spectra and ORD curves of ribonuclease and the ribonuclease-4-thiouridylic acid complex in the near u.v. and u.v. regions. ——, ribonuclease in 0·01 M acetate buffer solution plus 0·15 M KF (pH 5·6); – – –, the curves computed by summation of optical rotation or CD intensity of the constituents, ribonuclease and 4-thiouridylic acid; –·–·–, the experimental curves of the ribonuclease–thionucleotide complex in 0·01 M acetate buffer plus 0·15 M KF (pH 5·6). Insets represent the difference of ORD and CD between the computed and the experimental curves of the complex. (From Samejima *et al.*, 1969.)

little effect on the chemical shifts of the histidine C2-H resonance in the complex.

ORD and CD can also be used to gain information about possible conformational changes in the ligand on binding to ribonuclease (Samejima *et al.*, 1969). Using 2′(3′)4-thio-UMP these authors observed that binding caused a change in the sign of the CD spectrum of the ligand compared to that observed for the free nucleotide (Fig. 18). Comparison of the ORD and CD curves for the free thio-nucleotide with the difference ORD and CD curves for the bound thio-nucleotide shows that the longer wavelength (~340 nm) positive Cotton effect in the free nucleotide is converted to a negative Cotton effect (~330 nm) in the bound nucleotide (Fig. 18) and the positive CD band centred around 320 nm in the free nucleotide is converted to a negative CD band at 330 nm with a small positive band at 385 nm. Since both the ORD and CD effects observed on binding are in the region of the λ_{max} of the thio-nucleotide, they were ascribed to perturbations of the ligand. They concluded that binding "induced a new asymmetry in the nucleotide molecule, the reverse of the original one, which is caused by the attachment of the nucleotide to the polypeptide chain of the protein" (Samejima *et al.*, 1969). Although they did not discuss their results in terms of which particular conformation of the nucleotide was bound to the enzyme, on the assumption that the conformation of the free nucleotide was close to *anti* for which there is good experimental evidence (Emerson *et al.*, 1967; Miles *et al.*, 1969; Sundaralingam, 1969; Prestegard and Chan, 1969; Schleich *et al.*, 1972), one could interpret the change in asymmetry observed on binding as evidence for a conformation in the bound state closer to *syn*. Several authors have attempted to correlate observed ORD/CD spectra with geometry about the glycosidic bond (Emerson *et al.*, 1967; Miles *et al.*, 1969). Direct application of their results to 4-thio-UMP binding to RNase discussed above leads to the conclusion that 4-thio-UMP is in an *anti* conformation in solution but in a *syn* conformation when bound to the enzyme. However, since the "prediction rules" concerning ORD/CD patterns expected for different nucleotide conformations formulated by these latter authors are based on experiments with nucleotides of different structure and since the thio-nucleotide used in the binding experiment was a mixture of 2′ and 3′ isomers, which have different binding constants and possibly different binding geometries (see below), the direct application of these "prediction rules" is probably too simplistic an approach. In contrast to the results of Cathou *et al.* (1965), Deavin *et al.* (1966) and Simons (1971), all of whom studied cytidine or uridine nucleotides, Samejima *et al.* (1969) observed no changes in the CD spectra below 300 nm which could be attributed to a conformational change of the enzyme on the binding of 2′(3′)4-thio UMP.

Oshima and Imahori (1971) obtained CD difference spectra for the ribonuclease–2′-CMP complex, and concluded from the appearance of these spectra that the conformation of this nucleotide did not change substantially on binding. The suggestion that 2′-CMP might have a conformation about the glycosidic bond, close to *syn* in the bound state was made by Meadows *et al.* (1969), based on an anomalous change in chemical shift of the ribose H1′ of this nucleotide on binding, and on model-building studies. X-ray crystallographic studies have shown (Richards and Wyckoff, 1971) that the conformations of 2′-CMP and 3′-CMP do differ when bound to the enzyme, with the 2′ nucleotide being in a "*syn*-like" conformation.

Some recent information concerning conformational changes in RNase induced by ligand binding as studied by difference circular dichroism has been furnished by Simons (1971). She had shown previously that the positive extremum in the CD spectrum of RNase at 241 nm was highly sensitive to the environment of a buried tyrosine residue (Simons and Blout, 1968). The behaviour of this positive band was used to follow the binding of 3′-CMP. CD difference spectra of RNase-3′-CMP are shown in Fig. 19. A negative ellipticity difference was observed below 256 nm in acetate buffer but not in phosphate buffer suggesting that the specific nucleotide complex did not form in the presence of this competing ligand which was in vast excess. Similar difference spectra were recorded for 3′-UMP in acetate buffer but 3′-AMP had no effect. Figure 20 shows the CD spectra of RNase, 3′-CMP and their complex. Since 3′-CMP has negligible ellipicity at 241 nm, the variation of the ellipticity at this wavelength with 3′-CMP binding was assigned to chromophores on the enzyme. The ellipticity change observed was consistent with the interpretation that a tyrosine residue on the enzyme was becoming more

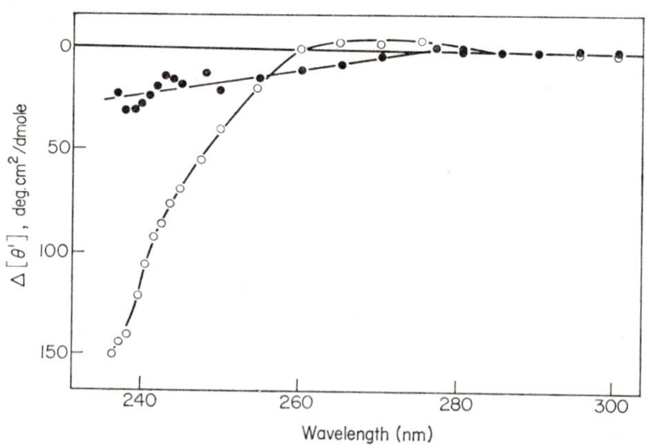

FIG. 19. Difference circular dichroism spectrum of ribonuclease-3′-CMP at pH 6·46, 5°C. ○——○, 0·1 M acetate; ●——●, 0·1 M phosphate. (From Simons, 1971.)

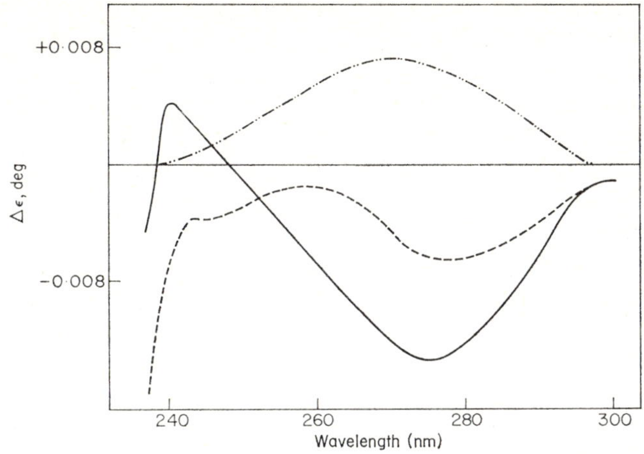

FIG. 20. CD spectra of ribonuclease (——), 3′-CMP (–. .–), and the ribonuclease-3′-CMP complex (– – – –) at equal concentrations in 0·1 M acetate buffer, pH 6·46, C = $5{\cdot}5 \times 10^{-5}$ M, 1 = 10·00 mm, 5°C. (From Simons, 1971.)

accessible to solvent, and since no tyrosine residues are in the immediate vicinity of the active site, the small conformational change observed on inhibitor binding must include more distant regions of the enzyme.

Yoshida *et al.* (1971) also recorded CD spectra for the binding of a 2′(3′)-CMP mixture to RNase and observed that the spectrum of the complex was not equivalent to the summation of the spectra for the free species; the actual difference spectrum was not calculated. Binding of GMP, however, resulted in a simple addition spectrum. The observation that purine nucleotides do not produce a CD difference spectrum suggests that binding of nucleotides to site B_2 does not produce an appreciable change in conformation.

V. Conformational Changes of Ribonuclease

Over the years evidence has accumulated from a variety of techniques to indicate that ribonuclease may undergo a pH dependent isomerisation or conformational change both in the free enzyme and in the enzyme nucleotide complex. Although a number of methods can detect the conformational change, few provide detailed information as to its precise nature. Hammes and co-workers have carried out the most extensive series of experiments on the phenomenon using chemical relaxation techniques. Cathou and Hammes (1964) made kinetic measurements on the binding of 3′-CMP to ribonuclease with the temperature jump technique and in addition to the relaxation time describing the bimolecular complexation reactions (τ_2) a second, concentration independent, relaxation time (τ_3) was observed. This was interpreted to reflect an isomerisation of the enzyme nucleotide complex which occurred subsequent to binding (see Fig. 11). The rate constant describing this conformational change was decreased when the solvent was changed to D_2O indicating that some form of proton transfer was involved. Later French and Hammes (1965) observed that ribonuclease itself could also undergo an isomerisation reaction characterised by a longer time constant (τ_1). This process is absent when 2′-CMP, 3′-CMP, cytidine or pyrophosphate are bound to the enzyme. It also does not occur if ribonuclease is carboxymethylated on His 119 or converted to RNase S; however the RNase-3′-CMP isomerisation (τ_3) does occur in RNase S. From the pH dependence of the isomerisation of the free enzyme, they concluded that a group with a $pK_a = 6{\cdot}1$ was being exposed in the process. The ΔH of ionisation for the group was approximately zero, leading to the suggestion that the isomerisation may represent the making and breaking of a hydrogen bond between a histidine and carboxyl group. No clear-cut chemical relationship between τ_1 and τ_3 was observed but Cathou and Hammes (1965) postulated that they both required the same group(s) on the enzyme. Hammes and Walz (1969) correlated this data with that obtained from nuclear magnetic resonance experiments (discussed below) and suggested that His 48 was the group involved. They postulated that His 48 had a $pK_a = 8{\cdot}5$ in the acid stable isomer of the free enzyme and in the initial complex. Isomerisation exposes the group and its pK_a becomes more normal (about 6) in both the free enzyme and in the nucleotide complex.

Temperature jump studies with substrates were begun by Erman and Hammes (1966a, b). The relaxation behaviour of C>p was found to be different to that observed with 3′-CMP. In the case of 3′-CMP, τ_1 increased as the concentration of 3′-CMP was increased until, at 25% saturation, the amplitude of the effect became too small to measure. With C>p however, τ_1 increased as the concentration of C>p was increased and finally levelled off. τ_1 was observable up to 95% saturation of the enzyme with C>p. The effects of CpC were found to be intermediate between these two extremes. These data suggested that the RNase-C>p complex could isomerise in the same way as free enzyme to some extent (characterised by a time constant τ_1) in addition to its much faster isomerisation process (characterised by τ_5) which is analogous to the τ_3 process seen with 3′-CMP. This ability to isomerise in a manner analogous to the free enzyme is a special feature of 2′ : 3′ cyclic phosphate substrates (see del Rosario and Hammes, 1970). Erman and Hammes (1966a, b) speculated that these enzyme nucleotide isomerisations (τ_3—3′-CMP, τ_5—C>p, τ_7—CpC) could reflect a conformational change at the active site which brings the catalytically important residues into the proper orientation.

The possible involvement of the 2′-hydroxyl group of the ribose in these isomerisations has been suggested by some recent experiments by Walz (1971). From a comparison of the pH dependence of the association rate constants for 3′-UMP and 2′-deoxy 3′-UMP (3′-dUMP) he deduced that the 2′-hydroxyl group must play a role in the formation of the initial complex. In addition the pH dependence of the relaxation kinetics indicated that isomerisation of the 3′-dUMP complex required only the involvement of two dissociating groups at the active site, whereas the isomerisation of the ribonuclease-3′-UMP complex depended on a dissociating group (perhaps His 48) extraneous to the active site. The formal mechanism for the isomerisation of the 3′-UMP complex is shown in Fig. 21. Knowing that a distant

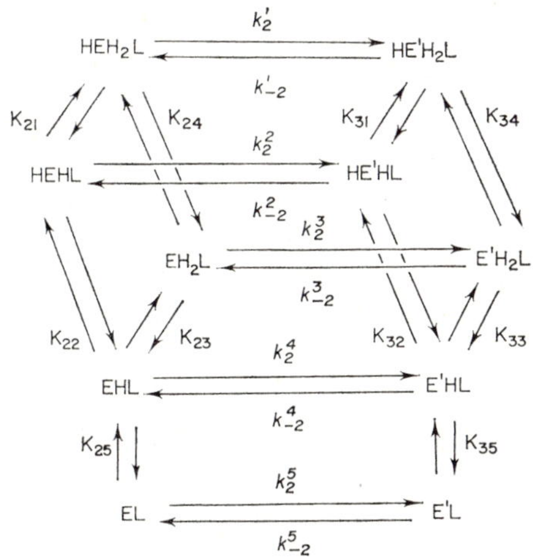

FIG. 21. A possible mechanism for the isomerisation of the RNase-3′-dUMP complex. Values for the rate and equilibrium constants are given in the original paper. (From Walz, 1971.)

group need not be implicated in the isomerisation of the ribonuclease-3′-dUMP complex he postulated that His 48 may only be involved in the isomerisation process if the nucleotide has a 2′-hydroxyl group. It is interesting to note that this bears some similarity to the proposal of Deavin *et al.* (1966) based on ORD measurements (see Section IVD).

At this stage it is necessary to point out that the temperature jump evidence concerning conformational changes presented above is, in general, indirect. In all but a few cases, pH indicators were employed in the experiments, so that proton uptake or release was measured. For example, the free enzyme isomerisation (τ_1) is characterised by a very fast liberation of protons, corresponding to a very fast equilibration between the two species E_1 and E_1H followed by a slow relaxation of the equilibrium $E_1H \rightleftharpoons E_2H$ concomitant with the uptake of protons. Hummel and Witzel (1966) observed that inhibitor binding above pH 5·5 was associated with an uptake of protons and below pH 5·5 with a release of protons. At pH 5·5 no uptake or release occurred. The extent to which different inhibitors caused the release or uptake of protons paralleled their K_i values and their ability to induce u.v. difference spectra. The phosphate group of the ligand did not seem to be involved since no effect occurred at pH 5·5 whatever the pK_a of the phosphate on the nucleotide. No effect was observed at any pH with A>p or other non-substrates, nor was the effect observable in CM-119-RNase. To the authors' knowledge, no correlation between these data and the temperature jump data has been attempted.

Some insight into the groups involved in the isomerisation of ribonuclease has come from the NMR experiments of Jardetzky and his colleagues. Meadows and Jardetzky (1968) observed that the resonance assigned to His 48 was broader than those of the other histidines in acetate buffer. In 0·2 M NaCl, the His 48 resonance broadens as the pH increases and is unobservable above pH 6. This was interpreted as exchange broadening of His 48, indicating that this residue may be undergoing an isomerisation at a rate that was dependent on pH and was affected by simple anions and by 3′-CMP. In addition, the small shift observed in His 48 on binding indicated that 3′-CMP was able to shift the position of the isomerisation equilibrium. Meadows *et al.* (1969) postulated that the effects of 3′-CMP on the linewidth and shift of His 48 could be rationalised by a closing of the active site cleft concomitant with inhibitor binding. This would enable Lys 41 to move closer to the phosphate ion in site p_1. The two "halves" of the ribonuclease molecule were pictured as moving closer together as units such that the only substantial rearrangement necessary was that about the "hinge" region which is located at the bottom of the active site cleft where the chain crosses three times between the two "halves" of the enzyme. Histidine 48 is located in this "hinge" region. These authors postulated that the two states between which His 48 was exchanging were: (a) A low pH state where histidine 48 is protonated, and thus has some aromatic character, leading to an interaction with tyrosine 25 which is almost directly above His 48 in the X-ray structure of RNase A. This stacking would account for the anomalous chemical shift of the C2 proton of histidine 48, since lying below tyrosine 25 it will experience an upfield shift due to the magnetic anisotropy of the aromatic ring. (b) At high pH, histidine 48 deprotonates, the interaction with tyrosine 25 is disrupted, and the shift of its C2 proton becomes more normal. (It is interesting to note in this regard that Donovan (1965) has observed the perturbation of one buried tyrosine residue by titration of a histidine residue using u.v. difference spectroscopy.) In RNase S tyrosine 25 and histidine 48 are further apart, the chemical shift of His 48 is more "normal" and no isomerisation is observed by temperature jump kinetics. All these facts are consistent with the above postulate. Rüterjans and Witzel (1969) have suggested a similar explanation of the behaviour of His 48, but involving its interaction with Asp 14 rather than Tyr 25. No distinction between these two possibilities can be made at the present time; it is likely that both contribute to some extent. The spin lattice relaxation times (T_1) of the C2 protons of the histidine residues in RNase have recently been measured (Benz *et al.*, 1972). Changes in these relaxation times on inhibitor binding reflect either changes in motion of the histidine residues or changes in the distance between the C2 protons and protons on other residues. A significant decrease in the T_1 for histidine 48 was observed on binding of 3′-CMP. This effect is consistent with a conformational change in this region of the molecule.

Markus *et al.* (1968) observed that the presence of 2′(3′)-CMP decreased the rate of subtilisin digestion of RNase to RNase S by 40% (Fig. 22). Since the Ala 20–Ser 21 bond cleaved by subtilisin is in the neighbourhood of His 48 in native RNase, this lends further support to the idea that a structural change occurs in this region of the molecule on inhibitor binding. Their results on inhibitor induced protection of RNase against trypsin and chymotrypsin digestion at high temperature (Fig. 23) suggest that this structural change may perhaps be more widespread (Markus *et al.*, 1968).

The closing of the active site cleft on nucleotide binding may be driven, at least in part, by an interaction of Lys 41 with the bound anion. Experimental support for the involvement of Lys 41 in a conformational change

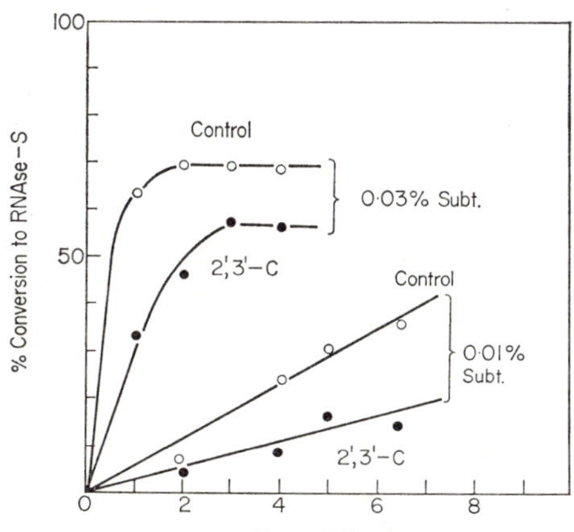

FIG. 22. Rate of ribonuclease-S formation at two concentrations of subtilisin. Ribonuclease A (0·93%) in 0·1 M acetate buffer, pH 5·5, digested at 2°C. Concentration of the inhibitor, 2′(3′)CMP: $1{\cdot}76 \times 10^{-2}$ M. Inhibitor to ribonuclease ratio 2·6. (From Markus *et al.*, 1968.)

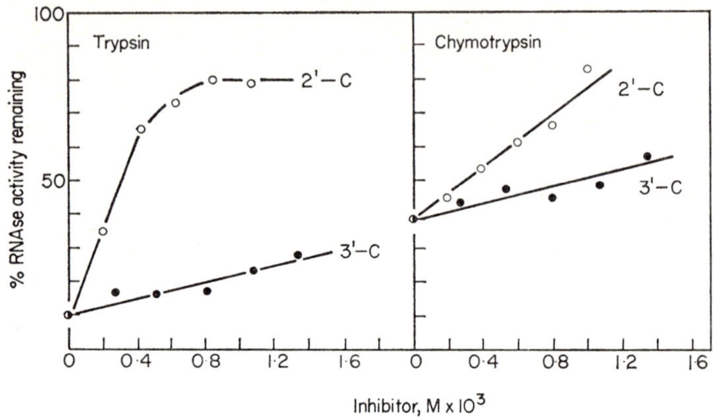

FIG. 23. Protection by 2′- and 3′-CMP against the tryptic and chymotryptic inactivation of ribonuclease A at 60° as a function of the concentration of inhibitor present during proteolysis. Trypsin: 1% ribonuclease in 0·01 M KCl, pH 5·5 digested for 15 min at 60°C by 0·02% trypsin. Chymotrypsin: as above, except 0·01 M $CaCl_2$, instead of KCl, and 0·05% chymotrypsin. (From Markus *et al.*, 1968.)

comes from the work of Ettinger and Hirs (1968). These authors made numerous physical measurements on DNP–Lys 41 RNase (Section IVA) in an attempt to gain some insight into the reason for the inactivity of this derivative. Crystallographic experiments have shown that the dinitrophenyl group of this derivative is somewhat buried in the enzyme and turned away from the active site (Allewell, 1969; Fung and Doscher, 1971). This has been confirmed by solvent perturbation and thermal transition measurements in solution (Ettinger and Hirs, 1968). In addition, these authors have shown that concomitant with the binding of 3′-CMP to 41 DNP–RNase, the DNP group becomes more exposed to solvent, as it does during the thermal transition.

Rüterjans and Witzel (1969) have put forward an alternative proposal for the ribonuclease isomerisation process seen in the temperature-jump experiments (Fig. 24). In this model they correlate the isomerisation with the breakdown of a diimidazole system (His 119-His 112) at the active site of RNase rather than with His 48. All species in the figure are expected to be in equilibrium with each other. A temperature jump will cause a perturbation of the equilibrium as shown in Fig. 24. The concentration of VII will increase after the temperature jump and the slow step in the process is the rupture of the diimidazole system V→VI. The explanation for the observed ΔH (ionisation)≈ 0 is that only the step involving V and VI is measured and not the proton uptake step VI→VII.

King and Bradbury (1971), on the basis of n.m.r. difference spectroscopic measurements on the C-4 protons of the histidine residue in ribonuclease, concluded that the broadening of the His 48 C-2 proton was not due to

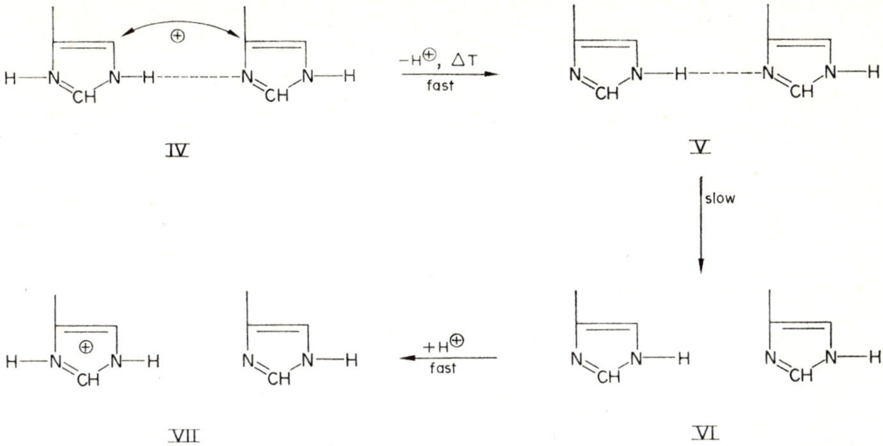

FIG. 24. A reaction scheme for the isomerisation process in the free enzyme as detected by temperature jump kinetics. (From Rüterjans and Witzel, 1969.)

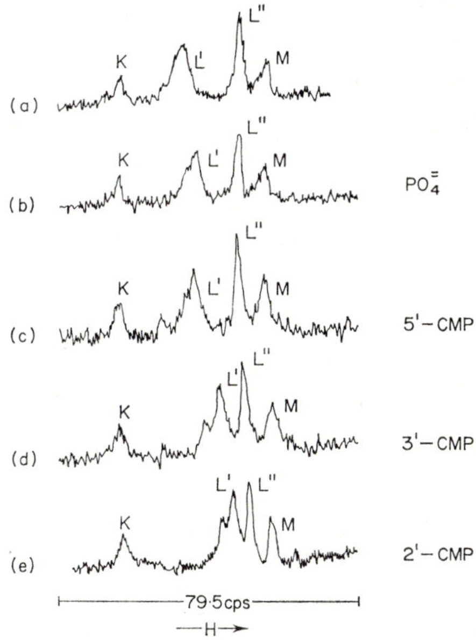

FIG. 25. ^{19}F-nuclear magnetic resonance spectra of triflyoroacetylated RNase S (4×10^{-3} M) in the presence of various inhibitors: (a) free RNase S; (b) $+10^{-2}$ M phosphate; (c) $+4 \times 10^{-3}$ M 5′-CMP; (d) $+4 \times 10^{-3}$ M c′CMP; (e) $+4 \times 10^{-3}$ M 2′-CMP. (From Huestis and Raftery, 1971.)

exchange broadening but rather due to its proximity to protons on other chemical groups, thereby potentially decreasing its spin-spin *and* spin-lattice relaxation time and broadening its resonance. Actual measurements of the spin-lattice relaxation times of the C-2 protons, however, indicate that, in acetate at least, the spin-lattice relaxation time of His 48 is quite comparable to those of the other histidines (Benz *et al.*, 1972). More detailed measurements will be needed to settle this point.

Huestis and Raftery (1971) have recently used ^{19}F-n.m.r. to monitor conformational changes in ribonuclease. Trifluoroacetylation of the two lysine residues on S-peptide yielded a derivative which, when combined with S-protein, had full enzymatic activity. The ^{19}F-n.m.r. spectrum of the reconstituted RNase S′ is shown in Fig. 25. From low to high field the —NH—CO—CF_3 resonances are assigned to trifluoroacetyl groups on the α-amino groups of Lys 1 (K), the ε-amino group of Lys 7 (L′) and the ε-amino group of single (L″) and doubly (M) substituted Lys 1. The effects of binding various ligands on the resonances is seen in Fig. 25. Phosphate and 5′-CMP caused identical 0·03 ppm shifts in L′ only. 2′ and 3′-CMP caused a 2–3 fold greater shift of L′ and caused the resonance to split into two components. In the complex with 3′-CMP, resonance L′ shows a pH dependent chemical shift with a $pK_a = 7{\cdot}3$. This is the pK_a of His 119 in the 3′-CMP complex. This was interpreted as evidence that the change on His 119 dominates the chemical shift of L′ only in the 3′-CMP complex. They concluded that inhibitor binding causes a small conformational change involving His 12 on the S-peptide which brings Lys 7 closer to His 119. The alternative hypothesis that, on binding of 3′-CMP, His 119 moves closer to Lys 7 is perhaps more consistent with the X-ray results discussed in Section II, which show His 119 stabilised in position III in the 3′-CMP complex. In this position His 119 would be extremely close to an ε-trifluoroacetyl group on Lys 7.

Using diazonium-1-H-tetrazole Horinishi *et al.* (1966) were able to modify the four histidine residues of ribonuclease in a stepwise manner (Fig. 26). Curve A in this figure represents the reaction in the presence of 3′-AMP but is identical to that for the native enzyme. Four plateau regions are observable, each corresponding to reaction of one of the histidine residues. Carrying out the reaction in the presence of 3′-CMP results in an increase in the reaction with the third histidine causing steps 2 and 3 to merge. In the presence of 3′-UMP either the second or fourth histidine residue reacts sooner causing two steps before the final plateau. 3′-AMP or 3′-GMP (curve A) had no effect. Since characterisation of the products at each plateau was not attempted, a detailed interpretation is not possible, but an *increase* in the rate of reaction of one histidine on inhibitor binding is strongly suggestive of a conformational change. The difference in the effects of 3′-CMP and 3′-UMP is striking but as yet inexplicable.

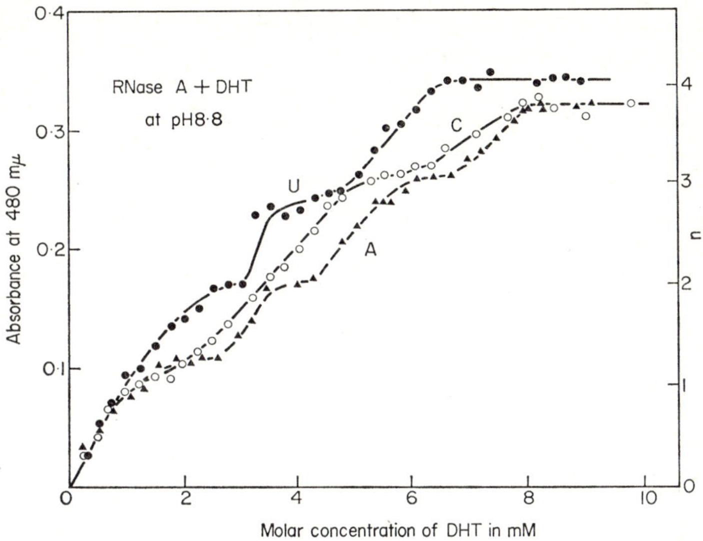

FIG. 26. Reaction curves of native RNase measured at pH 8·8 in the presence of adenosine-3′-phosphate (curve A), cytidine-3′-phosphate (curve C), and uridine-3′-phosphate (curve U). The curve obtained in the presence of guanosine-3′-phosphate and the curve obtained for native RNase were identical to curve A. The concentrations of RNase and nucleotides in the reaction mixtures were 4·9 and 72 μM respectively. (From Horinishi *et al.*, 1966.)

Further indirect evidence for a conformational change in the vicinity of His 48 can be found by comparing the results on the photo-oxidation of RNase obtained by Kenkare and Richards (1966) with those of Waku and Nakazawa (1965). The former authors showed that His 48 was resistant to photo-oxidation in the native enzyme. As mentioned in Section IVA, Waku carried out photo-oxidation experiments in the presence of 2′(3′)UMP or CMP and isolated a homogeneous material that had full activity but was lacking two histidines. They assumed His 12 and His 119 were intact, since the material was active, and that 48 and 105 had reacted. This means, however, that binding of inhibitors has resulted in His 48 becoming available for reaction. Without the characterisation of the reaction products however, this must remain speculative.

Nonnenmacher *et al.* (1971) carried out a deuterium–hydrogen exchange study of inhibitor induced conformational changes. They showed ribonuclease had four classes of exchangeable protons and the class involving instantaneously exchanged atoms was *increased* by about 10 hydrogens when 2′ or 3′-CMP bound. The number of exchanged hydrogens (in 24 hours) was not altered by 3′-CMP but was decreased by one half by 2′-CMP. They interpret this data as a loosening of the ribonuclease structure which correlated with the positive entropy change observed by Cathou and Hammes (1965).

Smith (1968) using electron paramagnetic resonance spectroscopy observed conformational changes in spin labelled ribonuclease derivatives on binding of nucleotides. Detailed analysis of the data is not possible since no direct characterisation of the site of attachment of the spin labels was made, but the data was consistent with a decrease in motional freedom of the labels on binding of RNA.

More ambitious attempts to fathom the nature of ribonuclease–nucleotide interactions by quantum chemical calculations have been made by Rein *et al.* (1971). Using a combination of molecular orbital and perturbation theories, they attempted to calculate the orientations of the His 12–His 119–3′-CMP groups in the nucleotide complex. Their calculations resulted in the prediction of an enlargement of the active site upon inhibitor binding, which is consistent with the positive entropy of binding observed by Cathou and Hammes (1965) and mentioned above.

The most recent experiment bearing on the conformational change question comes from Witzel's group (Rübsamen *et al.*, 1972). These authors have observed sigmoidal kinetics of C>p hydrolysis by ribonuclease when the assays were carried out above pH 7. A cyclic kinetic scheme which included two forms of the enzyme with different catalytic activities in a relatively slow pH-dependent equilibrium was used to explain the data. Further analysis must await the publication of the full report of these experiments.

One of the most controversial topics in discussions on ribonuclease conformational changes is the question of the involvement of tyrosine residues. Some of the conflicting reports on this subject have been mentioned in Section IV. Mathias *et al.* (1960), Findlay *et al.* (1961), Ross *et al.* (1962) and Hummel *et al.* (1961) suggested, on the basis of u.v. spectral measurements, that tyrosine may be perturbed on ligand binding. Irie and Sawada (1967), using the same technique, observed no u.v. difference spectra with 5,6 dihydro UMP binding nor with 4-thio-UMP binding which could be attributed to perturbation of groups on the enzyme. Cathou *et al.* (1965) presented evidence from ORD measurements that tyrosine residues were perturbed when 2′ or 3′-CMP bound to the enzyme. Samejima *et al.* (1969) using 2′(3′)4-thio-UMP could find no ORD/CD changes below 300 nm, indicating that protein residues were not perturbed. Simons (1971) correlated the change she observed in the CD spectrum of ribonuclease-3′-CMP complex with the exposure of one buried tyrosine. Hammes and Walz (1970) have been able to measure an isomerisation reaction of iodinated ribonuclease by temperature jump techniques observing changes in the absorption of the iodotyrosine residues and found that the time constant for this isomerisation was identical to that for the native enzyme. It is interesting to note that the three iodinated tyrosines (residues 73, 76 and 115) are in a region of the molecule quite far removed from the region of histidine 48, tyrosine 25 and the 20–21 peptide bond. The nearest of these three exposed tyrosines to the active site is residue 115; Friedman and Scheraga (1966) showed that binding of 3′-CMP did not block the iodination of this residue.

The recent results of Sawada (1969) point to a more direct involvement of tyrosine residues than any suggested hitherto. He found that ribonuclease was inactivated by irradiation with light at 330–365 nm if 4-thio-UMP was present. No inactivation occurred in the absence of nucleotide, nor in the presence of 4-thiouridine (except at very high concentrations). Inactivation by 4-thio-UMP and u.v. light was inhibited by the binding of 3′-UMP. The inactivation was notable at pH 5·6 but feeble at pH 3·6 and 7·6. The only modification was that of a single *tyrosine* residue. Its sequence position was not determined.

VI. Some Outstanding Problems in our Understanding of Ribonuclease–Nucleotide Interactions

In this section we shall consider observations which are not readily explicable at the present stage of our knowledge of this system, and where appropriate we shall speculate on possible interpretations.

A. DO THE CARBOXYMETHYL DERIVATIVES OF RIBONUCLEASE BIND NUCLEOTIDES?

It seems to be clear that the carboxymethyl-His 12 and carboxymethyl-His 119 derivatives of ribonuclease are enzymatically inactive (see Section IVA). However there is considerable disagreement in the literature as to the possibility that mononucleotides may still bind to these derivatives. Mathias, Rabin and their colleagues (Mathias *et al.*, 1960; Findlay *et al.*, 1961; Ross *et al.*, 1962) found that 2′-CMP and 3′-CMP generated u.v. difference spectra on addition to carboxymethyl-119 ribonuclease which were similar to those observed with the native enzyme. However, Hummel and co-workers (Hummel and Dreyer, 1962; Yang and Hummel, 1964) and Barnard and Ramel (1962) were unable to detect any interaction of 2′-CMP with carboxymethyl-119 ribonuclease or carboxymethyl-12 ribonuclease by gel filtration, u.v. difference spectroscopy or ultracentrifugation.

Recently Sawada and Irie (1969) re-examined this question, again using u.v. difference spectroscopy; their results are shown in Fig. 27. They calculated a binding constant for 3′-UMP to carboxymethyl-119 ribonuclease of $2{\cdot}5 \times 10^2\ \text{M}^{-1}$, compared to a value of $8{\cdot}5 \times 10^4\ \text{M}^{-1}$ for the complex with the native enzyme. The interaction with the carboxymethylated enzyme was too weak to be detected by gel filtration (in agreement with the work of Hummel and Dreyer, 1962; see Fig 28). Thus it appears that nucleotides may indeed

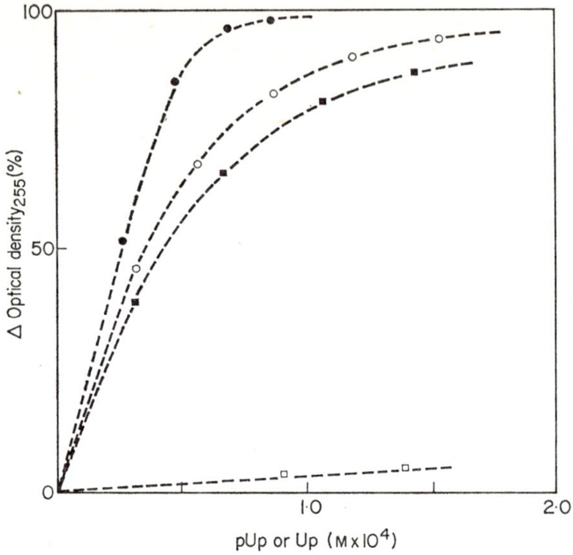

FIG. 27. Deviation of the observed absorbancy from that expected from summation of the constituent absorbances at 255 nm of the mixture of native RNase or 1-CM-119-RNase and increasing amounts of Up or pUp at pH 5·0 (0·01 M acetate) and 25°C. The concentrations of enzymes were $5{\cdot}6 \times 10^{-5}$ M for native RNase and $4{\cdot}0 \times 10^{-5}$ M for 1-CM-119-RNase. ----○----: Native RNase and Up; ----●----: Native RNase and pUp; ----□----: 1-CM-His-119-RNase and Up; ----■----: 1-CM-His-119-RNase and pUp. The theoretical curves (– – –) were drawn by taking the K_i values given in the text. The absorbancy difference was represented as percentage of the maximum difference ($\Delta\epsilon$). The $\Delta\epsilon$ values were 3000 for native RNase and both Up and pUp and 2400 for 1-CM-His-119-RNase and pUp. The value for 1-CM-His-119-RNase and Up was assumed to be 2400. (From Sawada and Irie, 1969.)

bind to carboxymethyl-His 119 ribonuclease, but some three hundred-fold more weakly than to the native enzyme. Interestingly, Sawada and Irie (1969) also showed that the binding of uridine 3′,5′-diphosphate was less affected by carboxymethylation of histidine 119; the binding constant fell from $6{\cdot}2 \times 10^{5}$ M^{-1} to $5{\cdot}5 \times 10^{4}$ M^{-1}. In this case the binding to the modified enzyme was strong enough to be detected by gel filtration (Fig. 28).

Examination of the model of the 3′-CMP-ribonuclease complex (Fig. 3) based on X-ray diffraction studies makes it clear that nucleotides could almost certainly not bind in the same way to the carboxymethyl-histidine 12 derivative, and indeed no report of nucleotide binding to this derivative has appeared. However, it is conceivable that binding to the carboxymethyl-119 derivative could occur if the carboxymethylated histidine side-chain could move far enough so as not to obstruct the phosphate binding site. There is chemical

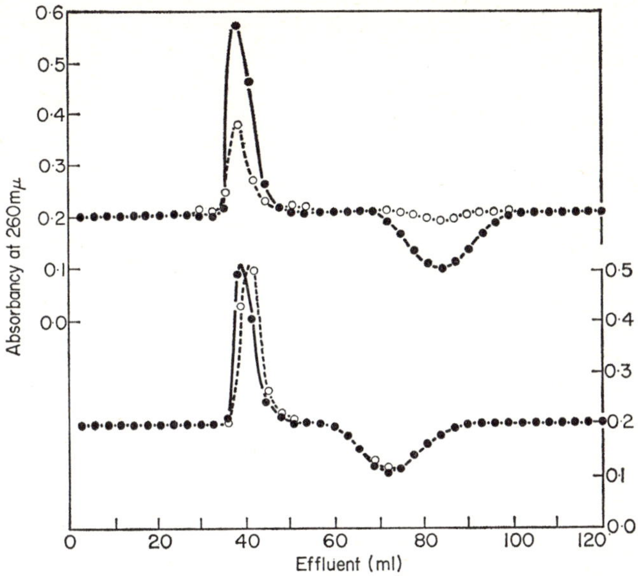

FIG. 28. Elution profile of the absorbancies at 260 nm on passage of RNase preparations through a column of Sephadex G-25. The columns (1·9 × 37 cm) were equilibrated and eluted at 4°C with solutions (0·01 M acetate, pH 5·6) of Up (above) and pUp (below), respectively. –.– native RNase; –○–, 1-CM-His-119-RNase. The concentrations of the nucleotides were $2{\cdot}0 \times 10^{-5}$ M and 3 mg each of the enzyme preparations were used in each case. (From Sawada and Irie, 1969.)

(see Section IVA) and n.m.r. (Meadows *et al.*, 1968) evidence for an interaction between the carboxymethyl group on histidine 119 and histidine 12, and this would have to be broken if the nucleotide was to bind in the same way as 3′-CMP. This, together with the replacement of a favourable interaction between the phosphate group and a positively charged histidine 119 by a repulsive interaction with the negatively charged carboxymethyl group would go some way towards explaining the dramatic decrease in binding constant.

Barnard (1969) has reported a personal communication from Kartha and Bello to the effect that binding of 5-iodouridine 5′ phosphate to crystals of carboxymethyl-119 ribonuclease A had been observed. The low resolution data available at that time indicated that the structure is closely similar to that of the native enzyme; it will be interesting to see if the high-resolution electron density map reveals a change in the orientation of histidine 119. The increased affinity of uridine 3′,5′ diphosphate both for the native and, particularly, for the carboxymethylated enzyme is harder to explain, since a separate binding site for the 5′ phosphate is not immediately apparent from the X-ray model.

B. WHAT IS THE ROLE OF THE 5′-LINKED NUCLEOTIDE?

The dramatic effect which the nature of the 5′ linked nucleotide can have on the rate of ribonuclease-catalysed hydrolysis of dinucleotides is illustrated in Table III, taken from the work of Witzel and Barnard (1962). While the

TABLE III

Kinetic constants at pH 7·0 (*imidazole buffer*), 26°
(K_m in $\text{M} \times 10^{-3}$, k_s in sec^{-1})
(*from Witzel and Barnard*, 1962)

Substrate	k_s	K_m	Substrate	k_s	K_m
CpA	3000	1·0	UpA	1200	1·9
CpG	500	3·0	UpG	—	—
CpC	240	4·0	UpC	40	3·0
CpU	27	3·7	UpU	11	3·7
Cp (cyclic)	5·5	3·3	Up (cyclic)	2·2	5·0
Cp-benzyl	2	3			
Cp-methyl	0·5 (from initial velocity, assuming K_m in the same range)				

K_m varies only by a factor of four, the turnover number (k_s) varies by a factor of over a hundred among the different dinucleotides. Wieker and Witzel (1967) also demonstrated that the rate of hydrolysis of cytidine and uridine 2′ : 3′ cyclic phosphates was substantially increased by the addition of adenine, adenosine or adenine nucleotides; again the major effect appeared to be on the catalytic rate, rather than on the binding. Witzel and his colleagues (Witzel and Barnard, 1962; Wieker and Witzel, 1967; Follman *et al.*, 1967), have proposed an explanation of these effects in which the rate-accelerating effect of the adenine base is attributed to a base-stacking interaction with the pyrimidine in the enzyme-substrate complex. This interaction is believed to orient the catalytically important groups (one of which, in Witzel's scheme, is the 2-carbonyl group of the pyrimidine) optimally for the reaction. This proposal is hard to reconcile with the structure of the UpcA complex deduced from crystallographic studies. In this structure (Fig. 4), the pyrimidine base is surrounded by amino acid side-chains in such a way that a stacking interaction would be highly improbable, and furthermore, the adenine base is bound to the site B_2 on the enzyme, in such a position that a direct interaction with the pyrimidine base can almost certainly be ruled out.

Thus, if the structure of the enzyme–UpcA complex is that of the enzyme–substrate complex—and there is no reason to believe that this is not so—an alternative explanation of the effects of adenine and guanine derivatives on the catalytic rate must be found. Richards and Wyckoff (1971) have suggested that these effects are due to the orientation imposed on histidine 119 by the binding of purines in the site B_2. As indicated in Section II, there is some uncertainty as to the orientation of the side-chain of histidine 119 in the absence of ligands. In the complex with 3′-CMP, it is found in position III (Fig. 3), while in the UpcA complex it is in position IV (Fig. 4). Binding of 3′-AMP (which stimulates cyclic phosphate hydrolysis) also moves histidine 119 to position IV. Pyrimidine nucleotides do not seem to bind to the site B_2; indeed in 2′deoxy TpT, there was little evidence for any binding of the 5′-linked nucleotide (Richards and Wyckoff, 1971). From this data, Richards and Wyckoff (1971) postulate that the accelerating effect of purine residues (as 5′ linked nucleosides or as activators of pyrimidine cyclic phosphate hydrolysis) is due to the stabilisation of histidine 119 in position IV which they produce; this position is taken to be the optimum one for catalysis. This attractive suggestion can obviously be tested by crystallographic studies of compounds such as UpcU and UpcG. It would be extremely interesting to be able to quantitate the effects on the catalytic rate of the precise orientation of histidine 119.

C. TO WHAT EXTENT ARE PURINE NUCLEOTIDES HYDROLYSED BY THE ENZYME?

The action of ribonuclease on poly A is extremely slow, and great care must obviously be taken to eliminate the possibility that it is due to a contaminating enzyme. Takemura *et al.* (1959), for example, showed that while commercial preparations of ribonuclease would hydrolyse ribo-apyrimidinic acids, purified ribonuclease A would not. Imura *et al.* (1965) carried out a careful investigation of the hydrolysis of poly A by carefully purified ribonuclease A. The patterns of oligonucleotides produced from poly A and poly U were similar, though a much higher enzyme concentration was needed for poly A digestion. Imura *et al.* (1965) found that the ratio V_{max}(poly U)/V_{max}(poly A) was 1300, while the ratio $V_{max}(U>p)/V_{max}(A>p)$ was over 500 000; thus the hydrolysis of poly A is relatively more efficient than that of A>p. This was attributed to the influence of interactions between poly A and multiple sites on the enzyme (compare the effects of adenine residues in site B_2 described above). Lane and Butler (1959) detected hydrolysis of ApAp by ribonuclease (at about 1/5000 the rate at which pyrimidine dinucleotides are hydrolysed), but could observe no hydrolysis of ApA. This indicates that multiple phosphate binding sites may also be involved. The activities of the enzyme against poly A and RNA declined in parallel on carboxymethylation of histidine 119, so there seems little doubt that the active site of ribonuclease can indeed catalyse the hydrolysis of poly A and adenine oligonucleotides, albeit very slowly. Levin (1962) observed cleavage of poly(8-azaguanylic acid) at about the same rate as poly A; the corresponding cyclic phosphate was not cleaved.

In contrast to the very slow hydrolysis of poly A, polyformycin (poly F; see Fig. 29) is hydrolysed at a rate comparable to that for poly C (Ward *et al.*, 1969; Ikehara *et al.*, 1969). Ward *et al.* (1969) showed that the nucleosides in poly F have the unusual *syn* conformation about the glycosidic bond,

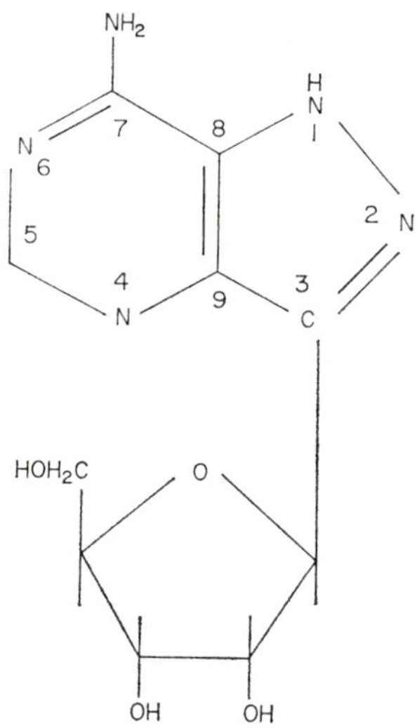

FIG. 29. Structure of formycin, 7-amino-3-(β-D-ribofuranosyl) pyrazolo-(4, 3-d) pyrimidine.

and that in this conformation formycin could fit in with the hydrogen-bonding scheme established for the pyrimidines (see Section II). Polylaurisin, which has a 7-keto group rather than the 7-amino group of polyformycin, was also found to have the *syn* conformation, and was also a good substrate. It is notable that all purine cyclic phosphates, even that of formycin, are extremely resistant to the action of ribonuclease. Ward *et al.* (1969) found that F>p was not detectably hydrolysed by the enzyme, and did not inhibit the hydrolysis of poly C, whereas Ikehara *et al.* (1969) were able to observe hydrolysis of F>p by using large quantities of enzyme. Two alternative explanations suggest themselves. First, additional nucleotides could be required, either to interact with other sites on the enzyme (cf. poly A), or to maintain the formycin or laurisin in the *syn* conformation. In the absence of these secondary interactions, purines appear to have extremely low affinity for site B_1. The other possibility, discussed in detail by Ward *et al.* (1969), is that the 2-carbonyl group of cytidine or uridine (or a stereochemically analogous one) is required for the hydrolysis of the cyclic phosphate but not for the transphosphorylation reaction. The importance of this carbonyl group in the mechanism has been advocated for some years by Witzel and his colleagues (e.g. Witzel, 1963). The arguments for and against this proposition are summarised by Richards and Wyckoff (1971); none of the arguments in its favour are conclusive, but neither has any clear disproof been presented. The differences between the two steps of the reaction (notably in rate, the transphosphorylation being 10–100-fold faster), are such as to make the proposal of Ward *et al.* (1969) worthy of further investigation, particularly in so far as there may be mechanistically significant differences between the structure of the dinucleotide complex and that of the cyclic phosphate.

D. HOW DOES RIBONUCLEASE CATALYSE OLIGONUCLEOTIDE SYNTHESIS?

The mere fact that ribonuclease does catalyse oligonucleotide synthesis—the reversal of the transphosphorylation reaction—is not surprising, and has been known for a number of years (Hepple *et al.*, 1955). However, Bernfield (1965, 1966; Bernfield and Rottman, 1967) have shown that a number of derivatives of ribonuclease which are inactive in the hydrolytic reaction seem to be able to catalyse the synthesis of oligonucleotides from pyrimidine cyclic phosphates. Thus while ribonuclease S-protein, carboxymethyl-His 119 ribonuclease and DNP-Lys 41 ribonuclease showed reduced synthetic activity compared to the native enzyme, their hydrolytic activity was reduced far more, so that a higher yield of oligonucleotide was obtained than with the native enzyme. Des (121–124) ribonuclease, however, was inactive in both the synthetic and hydrolytic reactions. The enhanced synthetic activity of S-protein was abolished by the addition of S-peptide. No explanation of these findings seems to be possible at the present time.

In addition to the examples discussed above, it should be pointed out that the simple hydrolysis of pyrimidine cyclic phosphates is not without its puzzling aspects. While the overall specificity pattern can, as mentioned in Section II, be satisfactorily accounted for by the X-ray structure, there is no straightforward explanation for the finding (Gassen and Witzel, 1967) that 4-thio-uridine 2′ : 3′ cyclic phosphate has a K_m of 1·9 mM and a k_s of 5·2 sec^{-1} under conditions where the values for U > p are 3·6 mM and 1·4 sec^{-1} respectively. A good deal more work will have to be done before our understanding of ribonuclease–nucleotide interactions is sufficient to explain the more subtle manifestations of the specificity of the enzyme.

References

Allewell, N. M. (1969). Ph.D. Thesis, Yale University.

Anderson, D. G., Hammes, G. G. and Walz, F. G. (1968). *Biochemistry* **7**, 1637.

Anfinsen, C. B. (1956). *J. biol. Chem.* **221**, 405.

Anfinsen, C. B., Sela, M. and Cooke, J. P. (1962). *J. biol. Chem.* **237**, 1825.
Anfinsen, C. B. and White, F. H. Jr. (1961). "The Enzymes", 2nd ed. **5**, 95.
Bald, R. W. and Holy, A. (1971). *Colln. Czech. chem. Commun.* **36**, 3657.
Barker, G. R., Hall, M. E. and Moss, R. J. (1961). *Biochim. biophys. Acta* **46**, 203.
Barnard, E. A. (1969). *A. Rev. Biochem.* **38**, 677.
Barnard, E. A. and Ramel, A. (1962). *Nature, Lond.* **195**, 243.
Barnard, E. A. and Stein, W. D. (1959). *J. molec Biol.* **1**, 339.
Bello, J. and Nowoswiat, E. F. (1965). *Biochim. biophys. Acta* **105**, 325.
Benz, F. W., Roberts, G. C. K., Feeney, J. and Ison, R. (1972). *Biochim. biophys. Acta* **278**, 233.
Bernfield, M. R. (1965). *J. biol. Chem.* **240**, 4735.
Bernfield, M. R. (1966). *J. biol. Chem.* **241**, 2014.
Bernfield, M. R. and Rottman, F. M. (1967). *J. biol. Chem.* **242**, 4134.
Borin, G., Toniolo, C., Moroder, E., Marchiori, F., Rocchi, R. and Scoffone, E. (1972). *Int. J. protein Res.* **4**, 37.
Bradbury, J. H. and Chapman, B. E. (1972). abstr. 418, *Abstr. commun. Meet. Fed. Eur. biochem. Soc. Amsterdam.*
Bradbury, J. H. and Scheraga, H. A. (1966). *J. Am. chem. Soc.* **88**, 4240.
Brown, D. M., Dekker, C. A. and Todd, A. R. (1952). *J. chem. Soc.* 2715.
Brown, D. M. and Todd, A. R. (1952). *J. chem. Soc.* 52.
Cathou, R. E. and Hammes, G. G. (1964). *J. Am. chem. Soc.* **86**, 3240.
Cathou, R. E. and Hammes, G. G. (1965). *J. Am. chem. Soc.* **87**, 4674.
Cathou, R. E., Hammes, G. G. and Schimmel, P. R. (1965). *Biochemistry* **4**, 2687.
Cohen, J. S. (1968). *Biochem. biophys. res. Commun.* **33**, 479.
Cooke, J. P. and Anfinsen, C. B. (1963). *J. biol. Chem.* **238**, 2034.
Crestfield, A. M. and Fruchter, R. G. (1967). *J. biol. Chem.* **242**, 3279.
Crestfield, A. M., Stein, W. H. and Moore, S. (1963a). *J. biol. Chem.* **238**, 2413.
Crestfield, A. M., Stein, W. H. and Moore, S. (1963b). *J. biol. Chem.* **238**, 2421.
Davis, F. F., Carlucci, A. F. and Roubein, I. F. (1959). *J. biol. Chem.* **234**, 1524.
Deavin, A., Fischer, R. C., Kemp, C. M., Mathias, A. P. and Rabin, B. R. (1968). *Eur. J. Biochem.* **7**, 21.
Deavin, A., Mathias, A. P. and Rabin, B. R. (1966). *Nature, Lond.* **252**, 55.
del Rossario, E. J. and Hammes, G. G. (1970). *J. Am. chem. Soc.* **92**, 1750.
Dickerson, R. E. and Geis, I. (1969). "The Structure and Action of Proteins", Harper, New York.
Donovan, J. W. (1965). *Biochemistry* **4**, 823.
Doscher, M. S. and Hirs, C. H. W. (1967). *Biochemistry* **6**, 304.
Doscher, M. S. and Richards, F. M. (1963). *J. biol. Chem.* **238**, 2399.
Eaker, D. (1961). *J. Polym. Sci.* **49**, 45.
Eckstein, F. (1968). *F.E.B.S. Letts.* **2**, 85.
Eckstein, F., Sanger, W. and Suck, D. (1972). *Biochem. biophys. Res. Commun.* **46**, 964.
Emerson, T. R., Swan, R. J. and Ulbricht, T. L. V. (1967). *Biochemistry* **6**, 843.
Erman, J. E. and Hammes, G. G. (1966a). *J. Am. chem. Soc.* **88**, 5607.
Erman, J. E. and Hammes, G. G. (1966b). *J. Am. chem. Soc.* **88**, 5614.
Ettinger, M. J. and Hirs, C. H. W. (1968). *Biochemistry* **7**, 3374.
Findlay, D., Herries, D. G., Mathias, A. P., Rabin, B. R. and Ross, C. A. (1961). *Nature, Lond.* **190**, 781.
Findlay, D., Herries, D. G., Mathias, A. P., Rabin, B. R. and Ross, C. A. (1962). *Biochem. J.* **85**, 152.
Finn, F. M., Dadok, J. and Bothner-By, A. A. (1972). *Biochemistry* **11**, 455.
Finn, F. M. and Hoffman, K. (1965). *J. Am. chem. Soc.* **87**, 645.

Finn, F. M. and Hoffman, K. (1967). *J. Am. Chem. Soc.* **89**, 5298.
Follman, H., Wieker, H. J. and Witzel, H. (1967). *Eur. J. Biochem.* **1**, 243.
French, T. C. and Hammes, G. G. (1965). *J. Am. Chem. Soc.* **87**, 4669.
Frensdorff, A., Wilchek, M. and Sela, M. (1967). *Eur. J. Biochem.* **1**, 281.
Friedman, M. E. and Scheraga, H. A. (1966). *Biochim. biophys. Acta* **128**, 576.
Fung, D. S. and Doscher, M. S. (1971). *Biochemistry*, **10**, 4099.
Gassen, H. G. and Witzel, H. (1965). *Biochim. biophys. Acta* **95**, 244.
Gassen, H. G. and Witzel, H. (1967). *Eur. J. Biochem.* **1**, 45.
Gilham, P. T. (1962). *J. Am. Chem. Soc.* **84**, 687.
Glick, D. M. and Barnard, E. A. (1970). *Biochim. biophys. Acta* **214**, 342.
Goren, H. J. and Barnard, E. A. (1970). *Biochemistry* **9**, 959.
Gross, E. and Witkop, B. (1967). *Biochemistry* **6**, 745.
Grundlach, H. G., Stein, W. H. and Moore, S. (1959). *J. biol. Chem.* **234**, 1754.
Hammes, G. G. (1968). *Acc. chem. Res.* **1**, 321.
Hammes, G. G. and Walz, F. G. Jr. (1969). *J. Am. chem. Soc.* **91**, 7179.
Hammes, G. G. and Walz, F. G. Jr. (1970). *Biochim. biophys. Acta* **198**, 604.
Heinrikson, R. L., Stein, W. H., Crestfield, A. M. and Moore, S. (1965). *J. biol. Chem.* **240**, 2921.
Heppel, L. A., Whitfield, P. R. and Markham, R. (1955). *Biochem. J.* **60**, 8.
Hirs, C. H. W. (1962). *Brookhaven Symp. Biol.* **15**, 154.
Hirs, C. H. W., Halmann, M and Kycia, J. H. (1965). *Arch. Biochem. Biophys.* **111**, 209.
Hoffman, K., Andreatta, R., Finn, F. M., Montibeller, J., Porcelli, G. and Quattrone A. J. (1971). *Bio-Organic Chem.* **1**, 66.
Hoffman, K., Finn, F. M., Montibeller, J. and Zanetti, G. (1966). *J. Am. chem. Soc.* **88**, 3633.
Holy, A. and Bald, R. W. (1971). *Colln. Czech. chem. Commun.* **36**, 2809.
Holy, A., Bald, R. W. and Sörm, F. (1972). *Colln. Czech. Chem. Commun.* **37**, 602.
Holy, A. and Sörm, F. (1968). *Biochim. biophys. Acta* **161**, 264.
Holy, A. and Sörm, F. (1969a). *Colln. Czech. chem. Commun.* **34**, 1929.
Holy, A. and Sörm, F. (1969b). *Colln. Czech. chem. Commun.* **34**, 3383.
Hooker, T. M. and Tanford, C. (1964). *J. Am. chem. Soc.* **86**, 4989.
Horinishi, H., Takenaka, O. and Shibuta, K. (1966). *Archs biochem. Biophys.* **113**, 371.
Huestis, W. H. and Raftery, M. A. (1971). *Biochemistry* **10**, 1181.
Hummel, J. P. and Dreyer, W. J. (1962). *Biochim. biophys. Acta* **63**, 530.
Hummel, J. P., Nelson, C. A. and Ver Ploeg, D. (1960). *Fed. Proc.* **19**, 344.
Hummel, J. P., Ver Ploeg, D. A. and Nelson, C. A. (1961). *J. biol. Chem.* **236**, 3168.
Hummel, J. P. and Witzel, H. (1966). *J. biol. Chem.* **241**, 1023.
Ikehara, M., Murao, K. and Nishimura, S. (1969). *Biochim. biophys. Acta* **182**, 276.
Imura, N., Irie, M. and Ukita, T. (1965). *J. Biochem* (*Tokyo*) **58**, 264.
Irie, M. (1965). *J. Biochem.* (*Tokyo*) **57**, 355.
Irie, M. (1968). *J. Biochem.* (*Tokyo*) **64**, 347.
Irie, M. and Sawada, F. (1967). *J. Biochem.* (*Tokyo*) **62**, 282.
Kartha, G., Bello, J. and Harker, D. (1972). *Nature, Lond.* **213**, 862.
Kenkare, U. W. and Richards, F. M. (1966). *J. biol. Chem.* **214**, 3197.
King, N. L. R. and Bradbury, J. H. (1971). *Nature, Lond.* **229**, 404.
Klee, W. A. and Richards, F. M. (1957). *J. biol. Chem.* **229**, 489.
Lamden, M. P., Mathias, A. P. and Rabin, B. R. (1962). *Biochem. biophys. Res. Commun.* **8**, 209.
Lane, B. G. and Butler, G. C. (1959). *Can. J. biochem. Physiol.* **37**, 1329.
Lee, G. C. Y. and Chan, S. I. (1971). *Biochem. biophys. Res. Commun.* **43**, 142.

Leinhard, G. E., Secemski, I. I., Keohler, K. A. and Lindquist, A. N. (1971). *Cold Spring Harb. Symp. quant. Biol.* **36**, 45.

Letters, R. and Michelson, A. M. (1962). *J. Chem. Soc.* 71.

Levin, D. H. (1962). *Biochim. biophys. Acta* **61**, 75.

Lin, M. C. (1970). *J. biol. Chem.* **245**, 6726.

Lin, M. C., Stein, W. H. and Moore, S. (1968). *J. biol. Chem.* **243**, 6167.

Mandel, H. G., Markham, R. and Matthews, R. E. F. (1957). *Biochim. biophys. Acta* **24**, 205.

Marchiori, F., Borin, G., Moroder, L., Rocchi, R. and Scoffone, E. (1972). *Biochim. biophys. Acta* **257**, 21.

Marfey, P. S., Uziel, M. and Little, J. (1965). *J. biol. Chem.* **240**, 3270.

Markham, R. and Smith, J. D. (1952). *Biochem. J.* **52**, 558.

Markus, G., Barnard, E. A., Castellani, B. A. and Saunders, D. (1968). *J. biol. Chem.* **243**, 4070.

Massoulie, J., Michelson, A. M. and Pochon, F. (1966). *Biochim. biophys. Acta* **114**, 16.

Mathias, A. P., Rabin, B. R. and Ross, C. A. (1960). *Biochem. biophys. Res. Commun.* **3**, 625.

Meadows, D. H. & Jardetzky, O. (1968). *Proc. natn. Acad. Sci. U.S.A.* **61**, 406.

Meadows, D. H., Jardetzky, O., Epand, R. M., Rüterjans, H. H. and Scherago, H. A. (1968). *Proc. natn. Acad. Sci. U.S.A.* **60**, 766.

Meadows, D. H., Markley, J. L., Cohen, J. S. and Jardetzky, O. *Proc. natn. Acad. Sci. U.S.A.* **58**, 1307.

Meadows, D. H., Roberts, G. C. K. and Jardetzky, O. (1969). *J. molec. Biol.* **45**, 491.

Michelson, A. M. (1962). *J. chem. Soc.* 979.

Michelson, A. M. (1963). "The Chemistry of Nucleosides and Nucleotides". Academic Press, New York and London.

Miles, D. W., Robins, M. J., Robins, R. K., Winkley, M. W. and Eyring, H. (1969). *J. Am. chem. Soc.* **91**, 831.

Mora, P. T. (1962). *J. biol. Chem.* **237**, 3210.

Moroder, L., Marcchiori, F., Rocchi, R., Fontana, A. and Scoffone, E. (1969). *J. Am. chem. Soc.* **91**, 3921.

Murdock, A. L., Grist, K. L. and Hirs, C. H. W. (1966). *Arch. biochem. Biophys.* **114**, 375.

Nelson, C. A., Hummel, J. P., Swenson, C. A. and Friedman, L. (1962). *J. biol. Chem.* **237**, 1575.

Nonnenmacher, G., Viala, E., Thiery, J. M. and Calvet, P. (1971). *Eur. J. Biochem.* **21**, 393.

Oshima, T. and Imahori, K. (1971). *J. Biochem. (Tokyo)* **70**, 193.

Potts, J. T., Young, M., Anfinsen, C. B. and Sadoval, A. (1964). *J. biol. Chem.* **239**, 3781.

Prestegard, J. H. and Chan, S. I. (1969). *J. Am. chem. Soc.* **91**, 2843.

Rein, R., Renugopalakrishnan, V. and Barnard, E. A. (1971). *Abstr. Y XIV B/3 First Euro. Biophys. Congr., Baden.*

Richards, F. M. (1958). *Proc. natn. Acad. Sci. U.S.A.* **44**, 162.

Richards, F. M. and Vithayathil, P. J. (1959). *J. biol. Chem.* **243**, 1459.

Richards, F. M. and Wyckoff, H. (1971). "The Enzymes" (ed. P. D. Boyer), 3rd ed., 4, Academic Press, New York and London.

Richards, F. M., Wyckoff, H. and Allewell, N. (1970). "The Neurosciences: Second Study Program" (ed. F. O. Schmitt). Rockefeller Press, New York.

Richards, F. M., Wyckoff, H., Carlson, W. D., Allewell, N., Lee, B. and Mitsui, Y. (1971). *Cold Spring Harbor Symp. quant. Biol.* **36,** 35.
Roberts, G. C. K., Dennis, E. A., Meadows, D. H., Cohen, J. S. and Jardetzky, O. (1969a). *Proc. natn. Acad. Sci. U.S.A.* **62,** 1151.
Roberts, G. C. K., Meadows, D. H. and Jardetzky, O. (1969b). *Biochemistry* **8,** 2053.
Rocchi, R., Marchiori, F., Moroder, L., Borin, G. & Scoffone, E. (1969). *J. Am. Chem. Soc.* **91,** 3927.
Ross, C. A., Mathias, A. P. and Rabin, B. R. (1962). *Biochem. J.* **85,** 145.
Rübsamen, H., Khandker, R., Witzel, H. (1972). *Abstr. 421,* Abstr. *Commun. Meet. Fed. Euro. Biochem. Soc.* **8** August.
Rüterjans, H. and Witzel, H. (1969). *Eur. J. Biochem.* **9,** 118.
Rüterjans, H., Haar, H., Maurer, W. and Thompson, T. C. (1971). *Abstr. Y IX B/5 First Eur. Biophys. Congr., Baden.*
Samejima, T., Kita, M., Saneyoshi, M. and Sawada F. (1969). *Biochim. biophys. Acta* **179,** 1.
Saunders, M., Wishnia, A. and Kirkwood, J. G. (1957). *J. Am. Chem. Soc.* **79,** 3289.
Sawada, F. (1969). *J. Biochem.* (*Tokyo*) **65,** 767.
Sawada, F. and Irie, M. (1969). *J. Biochem.* (*Tokyo*) **66,** 415.
Sawada, F. and Ishi, F. (1968). *J. Biochem.* (*Tokyo*) **64,** 161.
Scheraga, H. A. and Rupley, J. A. (1962). *Advan. Enzymol.* **24,** 161.
Schleich, T., Blackburn, B. J., Lapper, R. D. and Smith, I. C. P. (1972). *Biochemistry* **11,** 137.
Simons, E. R. (1917). *Biochim. biophys. Acta* **251,** 126.
Simons, E. R. and Blout, E. R. (1968). *J. biol. Chem.* **243,** 218.
Smith, I. C. P. (1968). *Biochemistry* **7,** 745.
Stark, C. R., Stein, W. H. and Moore, S. (1961). *J. biol. Chem.* **236,** 436.
Stein, W. D. and Barnard, E. A. (1959). *J. molec. Biol.* **1,** 330.
Sundaralingam, M. (1969). *Biopolymers* **7,** 821.
Takemura, S., Takazi, M., Miyazaki, M. and Egami, F. (1959). *J. Biochem.* (*Tokyo*) **46,** 1149.
Ukita, T., Hayatsu, H. and Waku, K. (1961a). *J. Biochem.* (*Tokyo*) **50,** 550.
Ukita, T. and Waku, K. (1964). *J. Biochem.* (*Tokyo*) **55,** 420.
Ukita, T., Waku, K., Irie, M. and Hoshino, O. (1961b). *J. Biochem.* (*Tokyo*) **50,** 405.
Usher, D. A. (1969). *Proc. natn. Acad. Sci. U.S.A.* **62,** 661.
Usher, D. A., Erenreich, E. S. and Eckstein, F. (1972). *Proc. natn. Acad. Sci. U.S.A.* **69,** 115.
Usher, D. A., Richardson, D. I. and Eckstein, F. (1970). *Nature, Lond.* **228,** 663.
Waku, K. and Nakazawa, Y. (1965). *J. Biochem.* (*Tokyo*) **57,** 578.
Walz, F. C. Jr. (1971). *Biochemistry* **10,** 2156.
Wang, J. H. (1968). *Science, N.Y.* **161,** 328.
Ward, D. C., Fuller, W. and Reich, E. (1969). *Proc. natn. Acad. Sci. U.S.A.* **62,** 581.
Weil, L. and Seibles, T. S. (1955). *Arch. biochem. Biophys.* **54,** 368.
Wellner, D., Silman, H. I. and Sela, M. (1963). *J. biol. Chem.* **238,** 1324.
Westheimer, F. H. (1968). *Acc. chem. Res.* **1,** 70.
Wieker, H. J. and Witzel, H. (1967). *Eur. J. Biochem.* **1,** 251.
Wigler, P. W. and Choi, H. U. (1964). *J. Am. chem. Soc.* **86,** 1636.
Winstead, J. A. and Wold, F. (1965). *J. biol. Chem.* **240,** 3694.
Witzel, H. (1963). *Prog. Nucleic Acid. Res.* **2,** 221.
Witzel, H. and Barnard, E. A. (1962). *Biochem. biophys. Res. Commun.* **7, 295.**

Wyckoff, H. W., Hardman, K. D., Allewell, N. M., Inagami, T., Johnson, L. N. and Richards, F. M. (1967). *J. biol. Chem.* **242**, 3984.

Wyckoff, H. W., Tsernoglou, D., Hanson, A. W., Knox, J. R., Lee, B. and Richards, F. M. (1970). *J. biol. Chem.* **245**, 305.

Yang, S. T. and Hummel, J. P. (1964). *J. biol. Chem.* **239**, 775.

Yoshida, N., Kuriyama, K., Iwata, T., and Otsuka, H. (1971). *Biochem. biophys. Res. Commun.* **43**, 954.

Notes added in proof

Since this review was completed, a number of important studies concerning the binding of nucleotides to ribonuclease have been reported. While extensive discussion of recent work is not possible, the essential conclusions are summarised below. (1) The n.m.r. experiments described in Section IV C showed that in the 3′CMP and 2′CMP-ribonuclease complexes both His 12 and His 119 are protonated, and the phosphate is in the dianionic state. This has now been shown *not* to be true for all nucleotides. Thus in the complexes with 5′AMP (P. H. Hoffner and J. H. Wang (1973), *Biochemistry*, **12**, 1608) and 3′UMP (J. H. Griffin, A. N. Schechter and J. S. Cohen, *Fifth International Conference on Magnetic Resonance in Biological Systems, New York*, 1972; *Ann. N.Y. Acad. Sci.*, in press, 1973) the pK of His 12 is increased by about the same amount as in the CMP complex, while that of His 119 is increased by only 0·3–0·4 units. The difference between 3′UMP and 3′CMP is particularly striking, since these two closely similar molecules are believed to occupy the same subsites, B_1, R_1, and p_1 (Section II A). (2) Griffin *et al.* (1972) have also studied the complex of UpcA with ribonuclease by n.m.r. and find that the pK value of *neither* His 12 nor His 119 is appreciably increased in the complex. This clearly argues for a somewhat different mode of binding of monoanionic dinucleotides, which are substrates, from that of dianionic mononucleotides, which are products (see Section II A, and p. 111). For a discussion of the implications of these results for our understanding of the mechanism of ribonuclease action, the original papers should be consulted. (3) The postulated interaction between His 119 and Asp 121 (see pp. 85, 101) has received some support from detailed analyses of the shape of the titration curve of His 119 as determined by n.m.r. spectroscopy (A. N. Schechter, D. H. Sachs, S. R. Heller, R. I. Shrager and J. S. Cohen (1972), *J. molec. Biol.* **71**, 39) and from n.m.r. studies of des(121–124)RNase (V. G. Sacharovsky, I. I. Chervin, G. I. Yakovlev, S. M. Dudkin, M. Y. Karpeisky, S. V. Shliapnikov and V. F. Bystrov (1973), *FEBS Letters* **33**, 323). The latter experiments showed that the pK_a of His 119 is markedly higher in des(121-124)RNase than in the native enzyme; the shape of its titration curve is also substantially altered. (4) R. Hayashi, S. Moore and R. B. Merrifield (1973, *J. biol. Chem.* **248**, 3889) report that the addition of a synthetic peptide (116–124) to RNase (1–114) or RNase (1–115), prepared by selective proteolytic cleavage, led to the regain of about 50% of the enzyme activity of the native molecule. This opens the way to detailed studies of the role of the amino acids in this region in activity and nucleotide binding, in a similar way to the experiments with synthetic S-peptides described in Section IV A. These investigators have already shown in this way that Tyr 115 is not essential for activity (cf. p. 126).

Acknowledgements

We are particularly grateful to Professor F. M. Richards for supplying us with the originals of the stereo diagrams.

CHAPTER 21

The Measurement of Radiation-induced Strand Breaks in the DNA of Mammalian Cells

M. G. Ormerod

Chester Beatty Research Institute, Institute of Cancer Research, Royal Cancer Hospital, Surrey, England

I. Introduction

Zone centrifugation of DNA in a density gradient of sucrose is widely used for the measurement of the molecular weight of DNA molecules, and, in particular, the method has been used to compare the molecular weights of DNA molecules isolated from bacteriophages (e.g. see Burgi and Hershey, 1963; Leighton and Rubenstein, 1969). When DNA is isolated from cells by conventional methods, extensive degradation caused by mechanical shear is unavoidable. To prevent this, McGrath and Williams (1966) developed a method in which suitably pre-treated bacteria were lysed directly on top of a sucrose gradient. The rate of sedimentation was then measured without any handling of a DNA solution. This technique has since been widely used to study the formation and repair of strand breaks introduced into the DNA of cells by high energy radiation. Most workers have employed alkaline gradients which denature DNA and thereby expose single-strand breaks.

Alkaline sucrose gradients sedimentation can be used to study any process which either

a. introduces strand breaks into DNA by direct chemical action
b. produces alkali-labile bonds, or
c. induces endonuclease attack.

Presently available techniques do not allow one to distinguish between a and b. Endonuclease attack can usually be observed by post-treatment breakdown of the cellular DNA although this effect is sometimes obscured by a rapid re-joining of strand breaks.

Neutral sucrose gradients have not been used as extensively, partly because it is more difficult to obtain uncontaminated DNA of high molecular weight and partly because the introduction of double-strand breaks is a less common phenomenon.

In this review, I have concentrated on the effects of high energy radiation, although many of the points raised are of a general nature which are applicable to any study of high molecular weight DNA on a sucrose gradient. In particular I have emphasised the importance of correct quantitative analysis of the data.

II. Experimental Techniques

Basically the experiment proceeds as follows. A sucrose gradient is formed in a centrifuge tube. A solution of lysing agent is layered on the top of the gradient followed by a cell suspension and the gradient is then centrifuged in a swinging bucket rotor for a fixed time and speed. Finally, the bottom of the tube is pierced, fractions collected and their DNA content estimated.

The gradient of sucrose is usually linear and is formed using an apparatus similar to that described by Britten and Roberts (1960). Burgi and Hershey (1963) have demonstrated that the distance sedimented through a 5%–20% gradient of sucrose is proportional to the sedimentation rate of a DNA molecule. Ideally, if the gradient conditions are changed, care should be taken to ensure that the gradients are isokinetic (i.e. the increasing centrifugal force down the centrifuge tube is balanced by the increase in viscosity and density of the gradient). This point has been discussed in detail by Noll (1967).

The sucrose gradient usually contains NaCl (between 0·1 and 2 M) and for the study of native DNA, a suitable buffer. For the study of denatured DNA, alkaline gradients are used. The pH of the gradients can either be adjusted to a given pH (usually 12) with NaOH or the normality with respect to NaOH can be fixed (usually 0·1 N or above). Sucrose has a mild buffering action and in the case of 0·1 N NaOH the 5% and 20% sucrose solutions will have pHs of 12·4 and 11·8 respectively.

Some workers also include a low concentration of EDTA (e.g. 3 mM, Elkind and Kampfer, 1970). The addition of other additives such as 6 M guanidine chloride (Moroson and Furlan, 1970) or 0·06 M p-amino-salicylic acid (Humphrey *et al.*, 1968) appear to be unnecessary since most workers obtain reproducible results without them.

The cells are pipetted into a lysing solution previously layered on to the gradient. The volume of this solution should not be greater than the volume of one fraction.

On alkaline gradients, a variety of lysing solutions have been used. The simplest utilised the gradient itself. The cells were layered on to a solution of 2% sucrose through which they settled; on contacting the alkaline gradient the cells lysed (McBurney *et al.*, 1971). In this laboratory we have found that, although the precise composition of the lysing solution is unimportant, the inclusion of EDTA (we use 0·02 M) gives more reproducible results.

On neutral gradients, it is necessary to use a detergent both to lyse the cells and to deproteinise the DNA. This will be discussed in more detail in a later section.

It is important that the cell number should be such that the total quantity of DNA on top of the gradient does not exceed 1 μg ml^{-1}. Increasing the quantity of DNA leads to distorted sedimentation profiles (Burgi and Hershey, 1963; Dean *et al.*, 1969). When alkaline gradients are being used, it should be noted that alkali slowly degrades DNA (Lett *et al.*, 1970; Elkind, 1971). The degradation is accelerated by light (Elkind, 1971), and if centrifuge tubes are not spun immediately after layering the cells, they should be stored in the dark.

Because of the small quantities used in these experiments, the DNA in the cells used will have to be radioactively labelled—usually with ^{3}H- or ^{14}C-thymidine. It is normal practice to pierce the bottom of the centrifuge tube and to collect a suitable number of drops on to numbered filter paper discs or filter paper strips. The discs are dried, washed in ice-cold 10% trichlor-acetic acid followed by ethanol and then acetone. After drying again the discs are placed in bottles containing a suitable scintillant and counted in a liquid scintillation counter. If a detergent is used as a lysing agent, the last few drops will be appreciably smaller. This can be quantitated by running a mock experiment in which all the solutions contain 1 M $CuSO_4$. The fractions are collected in tubes, made up to 2 ml with water and the absorbance read at 750 mμ, this being proportional to the volume of each fraction.

III. Analysis of Data

Doty *et al.* (1958) first proposed the relationship

$$S = b\,\text{M}^{a}$$

where S is the sedimentation coefficient and M the molecular weight of the DNA; a and b are constants. This has been confirmed by Burgi and Hershey (1963) and Studier (1965) over a molecular weight range extending up to $1{\cdot}2 \times 10^8$ daltons for native DNA and $0{\cdot}6 \times 10^7$ daltons for denatured DNA. The relation between S and the distance sedimented by a DNA molecule (D) is given by (Burgi and Hershey, 1963)

$$S = c\,\frac{\mathrm{D}}{\omega^2 t}$$

where ω is the angular velocity of the rotor which is spun for time t; c is a constant. The two equations can be combined to give

$$\mathrm{M} = K\left(\frac{\mathrm{D}}{\omega^2 t}\right)^{1/a}$$

The values of "a" which have been measured are

for native DNA:	0·346 by Studier (1965)
	0·35 by Burgi and Hershey (1963)
for denatured DNA:	0·400 by Studier (1965)
	0·38 by Abelson and Thomas (1966).

In our work we have used the values obtained by Studier. (It should be noted that the above relationship has only been confirmed for native DNA of molecular weights up to $1{\cdot}3 \times 10^8$ daltons and denatured DNA of weights up to $0{\cdot}65 \times 10^7$ daltons. It is assumed that the same equation can be extrapolated to higher molecular weights but this cannot be confirmed since there is no source of higher molecular weight DNA of known length). The distance sedimented will depend upon the viscosity, density, temperature and, in the case of denatured DNA, the ionic strength of the gradient. This means that the constant K must be determined separately for every type of gradient used. Phage DNA of known molecular weight can be used as a standard; the methods for preparing radioactive DNA from *E. coli* phages have been described by Abelson and Thomas (1966). For calibration of alkaline gradients, we have found that T2 phage can be directly lysed on top of the gradient in 0·5 N NaOH/0·02 M EDTA.

D is calculated from the relationship

$$D = \frac{Li}{N}$$

where N is the number of fractions in the linear part of the centrifuge tube which is of length L and i is the fraction number counted from the top of the tube. If detergent was used to lyse the cells, N and i must be corrected to allow for the smaller size of the topmost fractions.

ω And t are measured using a stopwatch and the tachometer on the centrifuge. The total centrifugal force exerted during running up and slowing down of the centrifuge rotor should also be estimated and added to the value of $\omega^2 t$.

If DNA is carefully isolated from a phage, all the molecules have the same molecular weight (i.e. the molecular weight distribution in monodisperse). The meaning of any molecular weight measurement is obvious. However, the DNA isolated from a cell has a polydisperse molecular weight distribution and a measurement of the molecular weight of the peak fraction is of limited value. Correctly analysed, a sucrose gradient experiment should yield both the average molecular weight and the molecular weight distribution.

The ensuing discussion will consider the number of nucleotides per molecule (u) so that

$$\text{M} = wu$$

where w is the average molecular weight of one nucleotide unit.

The number average molecular weight $= \text{M}_n = u_1 w$

The weight average molecular weight $= \text{M}_w = u_2 w$

$$\text{By definition, } u_1 = \frac{\Sigma nu}{\Sigma n}$$

$$\text{and } u_2 = \frac{\Sigma nu^2}{\Sigma nu}$$

when n is the number of molecules containing u units and the summations are taken over values of u from 0 to ∞. The range of values of "u" collected in the ith fraction from the gradient is Δu_i and the average value of u is u_i. The amount of radioactivity in that fraction (c_i) equals (average length of label per molecule) $\times$ (number of molecules of length u_i) $\times$ Δu_i. So that

$$c_i = f u_{\text{lab}}\, n_i \Delta u_i$$

where f is the specific activity.

If the DNA molecules are labelled throughout their length, $u_{\text{lab}} = u_i$ and $c_i = f n_i\, u_i \Delta u_i$ so that $n_i = c_i / f u_i \Delta u_i$. A plot of $c_i / u_i \Delta u_i$ against u_i will demonstrate the molecular weight distribution. In terms of molecular weight $c_i / \text{M}_i \Delta \text{M}_i$ is plotted against M_i.

M_n and M_w can be calculated from the expression $\text{M}_n = \Sigma c / \Sigma c/\text{M}$ and $\text{M}_w = \Sigma c\text{M} / \Sigma c$. These values will always be inaccurate; using experimental data the summation can only be carried out over finite limits. Furthermore, inaccuracies in the amount of activity collected in fractions of low molecular weight will have a disproportionate effect on the value of M_n; while inaccuracies in high molecular weight fractions will distort the value of M_w. It is preferable to estimate M_n and M_w from the known characteristics of the molecular weight distribution. This is demonstrated below for a particular case.

High energy radiation introduces strand breaks into the DNA of cells. Since the breaks are introduced at random, the resulting molecular weight distribution is random and is given by the expression

$$n_i = \sum \frac{u_i n_i}{u_1^2} \exp(-u_i/u_1)$$

(Charlesby, 1954; Lehamnn and Ormerod, 1970a), so that

$$\frac{c_i}{u_i \Delta u_i} = \sum \frac{u_i n_i}{u_1^2} \exp(-u_i/u_1)$$

Therefore a plot of log $(c_i/u_i\Delta u_i)$ against u_i will give a straight line of slope $-1/u_1$. We normally plot log $(c_i/M_i\Delta M_i)$ versus M_i and read off M_n from the plot. This plot not only gives a reasonably accurate value of M_n but also demonstrates that the molecular weight distribution is indeed random. An example of this type of plot is shown in Fig. 1. In the case of a random molecular weight distribution, $M_w = 2M_n$.

It can be shown that the molecular weight of the peak fraction on the gradient (M_p) is related to the number average molecular weight by the expression:

$$\frac{M_p}{M_n} = 2 - a$$

(where "a" is the constant in the equation $M = K(D/\omega^2 t)^{1/a}$). (Lehmann and Ormerod, 1970a).

For DNA sedimented on an alkaline gradient, $a = 0{\cdot}40$ and $M_p/M_n = 1{\cdot}6$; for native DNA, $a = 0{\cdot}35$ and $M_p/M_n = 1{\cdot}65$. This gives a quick method of estimating M_n for a sedimentation profile if it is known that the molecular weight distribution is random.

If radiation is used to shear the DNA and the breaks are introduced at random as a single event, then

$$\frac{1}{M_n} - \frac{1}{M_0} = \text{constant} \times r$$

where M_0 is the molecular weight of the DNA extracted from an unirradiated cell and r is the radiation dose. A plot of $1/M_n$ against radiation dose should be a straight line the slope of which will give the break efficiency. Extrapolation to zero dose should yield an estimate of the M_0.

These comments are only applicable if the DNA is fully labelled. If this is not the case (for example, if the cells were pulse-labelled with tritiated thymidine) then the distribution of radioactivity will not represent the molecular distribution. This situation has been analysed in detail by Lehmann and Ormerod (1969, 1970a). They have shown that, for pulse-labelled, irradiated cells on alkaline gradients, M_p/M_n can vary from 0·6 to 1·6 depending on the length of label and M_n. The data obtained can be used to measure the replication rate of the DNA (Lehmann and Ormerod, 1970a). This effect

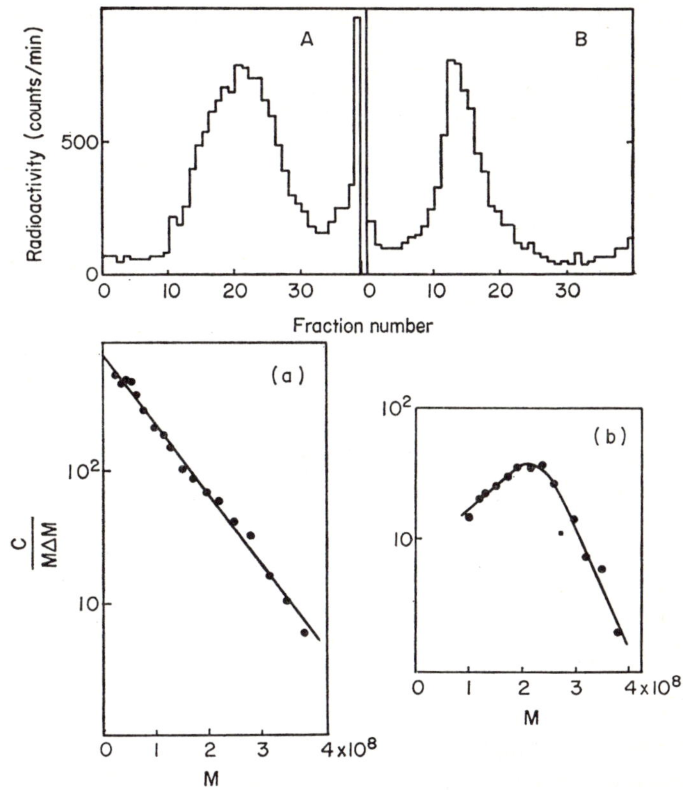

FIG. 1. Alkaline sucrose gradient profiles and associated C/MΔM plots. Gradient: 4·9 ml of 5–20% (w/v) sucrose, 0·1 M NaOH, 0·9 M NaCl. Lysing solution: 0·1 ml of 0·1% Nonidet P40 (from Shell Chemicals U.K. Ltd.), 0·1 M NaOH, 0·02 M EDTA. L5178Y cells labelled with ^{3}H-thymidine (10^4 cells were layered onto each gradient). The gradients were spun on a Beckman L2 centrifuge in a SW50 head for 1 hour at 35 000 rev/min. A. Irradiated with 30 krads of X-rays and then incubated at 37°C for 60 min. B. Irradiated with 10 krads of X-rays and then incubated for 40 min. (a) and (b) are the C/MΔM plots associated with A and B respectively. Plot (a) demonstrates that the DNA in profile A has a random molecular weight distribution while plot (b) shows the non-random distribution in profile B. The latter effect is due to a sedimentation anomaly which is discussed later in the text. Fraction 1 is at the bottom of the centrifuge tube. (Taken from Ormerod and Stevens, 1971.)

complicates the study of newly replicated DNA and great care must be taken when analysing the results. Lehmann (1972) has discussed this in detail in the case of u.v. irradiated cells.

Analysis of the data will only be meaningful if the DNA on the gradient is uncontaminated and sediments according to theory. In order to establish that the DNA released from cells behaves as free molecules, we have used several criteria (Lehmann and Ormerod, 1970b). These are:

1. there should be no contamination of the DNA by RNA, lipid or protein as detected by double-labelling;
2. within experimental error, all the DNA should be recovered from the gradient;
3. the measured molecular weight should be independent of the ionic strength of the gradient.

If one is studying DNA from irradiated cells (as was the case in our experiments) then one can add the following two criteria:

4. the molecular weight distribution should be random;
5. the number of breaks produced (as measured by $1/M_n$) should be linear with radiation dose.

As will be discussed later in the paper, we have been unable to isolate unsheared free native DNA from unirradiated cells, bacterial or mammalian. Similarly, free denatured DNA cannot be isolated from unirradiated mammalian cells without some degradation of the DNA although determination of the molecular weight of denatured DNA from bacteria and protoza may be possible.

IV. Experimental Studies of Native DNA

Denatured DNA of a high molecular weight from mammalian cells is usually studied on alkaline sucrose gradients. The alkali lyses the cells, denatures DNA, hydrolyses RNA and also appears to de-proteinise the DNA. In the study of native DNA, one of the major problems is to ensure that the DNA is freed of both RNA and protein. It is important to ensure that all the criteria listed in the analysis section are fulfilled. There have been few published attempts to isolate native DNA on sucrose gradients (Veatch and Okada, 1969; Terasima and Tsuboi, 1969) and the only acceptable study appears to be our own which is described below (Lehmann and Ormerod, 1970b).

In order to lyse the cells and de-proteinise the DNA we have used a detergent solution. Satisfactory results were obtained either with sodium dodecyl sulphate or sodium dodecyl sarkosinate. A chelating agent (sodium citrate) was included in the gradient but it was not needed in the lysing solution.

Nucleases from the murine lymphoma cell under study were inhibited by the detergents. This may not always hold. When we attempted to repeat our experiments using a protozoan, *Tetrahymena pyriformis*, we were unable to inhibit completely the action of an endonuclease (White, 1971).

The DNA isolated from unirradiated L5178Y cells was in the form of a rapidly sedimenting complex which contained lipid (Ormerod and Lehmann, 1971); free DNA could be isolated from the cell only after the DNA had been sheared by X-irradiation. The pattern of release of free DNA from the complex by X-rays has been described by Ormerod and Lehmann (1971b). On sucrose gradients, approximately 80% of the DNA was released by 30 krads of radiation; complete release was obtained after 100 krads. We therefore measured the efficiency of double-strand production at radiation doses in excess of 30 krads.

If the speed of centrifugation did not exceed 10^4 rev/min, all the criteria listed in the Analysis section were fulfilled in the range of radiation dose covered (30–400 krad) (Lehmann and Ormerod, 1970b). Typical profiles together with the associated linear $C/M\Delta M$ plots are shown in Fig. 2. From the data obtained with gradients of ionic strength varying from 0·1 to 2, it was calculated that one double-strand break is formed by the absorption of 2900 ± 400 eV of radiation energy by the DNA—this is equivalent to 1 break per krad per 3×10^{10} daltons. From the extrapolation of a plot of $1/M_n$ versus dose to zero dose, it is estimated that the molecular weight of the DNA in the unirradiated cell is at least 3×10^9 daltons so that it may be concluded that the DNA in the chromosome consists of not more than 30 molecules and may be a single molecule of molecular weight 10^{11} (Lehmann and Omerod, 1970b).

If M_n was 10^8 daltons or less, the shape of the sedimentation profile was independent of rotor speed. However, if M_n was greater than 10^8 daltons, this was no longer true. Increasing the rotor speed to 2×10^4 rev/min or above resulted in sharp non-random profiles with a peak molecular weight of about 2×10^8 daltons. If M_n was estimated by the summation method, a plot of $1/M_n$ against radiation dose was non-linear (Lehmann and Ormerod, unpublished work). These observations were not altered by using different detergents or changing the ionic strength of the gradient. We conclude either that this effect is caused by minor contamination of the DNA by other cellular components or that high molecular weight DNA sediments anomalously at high rotor speeds. The latter explanation is probably correct since the same effect has been observed with bacterial DNA and with high molecular weight denatured DNA (see below).

When irradiated cells were incubated in growth medium at 37°C we were unable to observe any evidence for rejoining of the double-strand breaks after doses of 20 krads or more (Lehmann and Ormerod, 1970b). Under these conditions single-strand breaks are rapidly rejoined but it must be emphasised this radiation dose is nearly 200 times in excess of the radiation doses used to study cell survival.

The behaviour of the fast-sedimenting complex from unirradiated cells suggests that the DNA is attached to the nuclear membrane by a detergent-stable bond (Ormerod and Lehmann, 1971a). The complex can be observed in a sedimentation experiment in three ways:

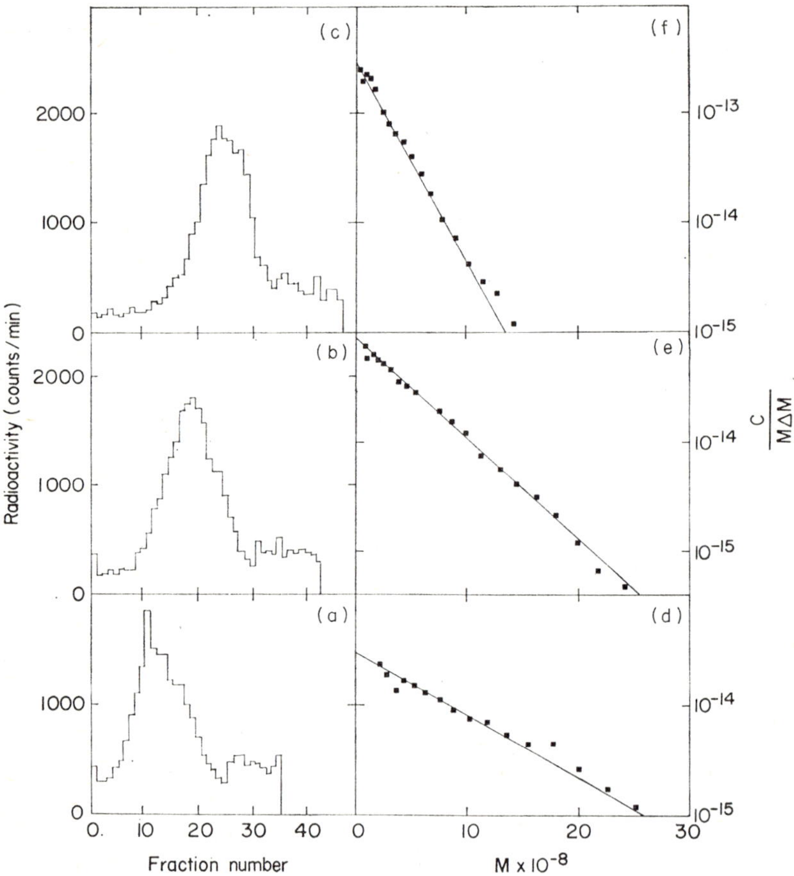

FIG. 2. Sedimentation profiles from native DNA. Gradient: 4·9 ml of 5–20% sucrose, 2 M NaCl, 0·01 M sodium citrate (pH 9). Lysing solution: 0·1 ml of 0·2% sodium dodecyl sarkosinate, 0·08% sodium deoxycholate, 2 M NaCl, 0·01 M sodium citrate, 0·02 M EDTA (pH 9). L5178Y cells labelled with ^{3}H-thymidine. Gradients were spun at 8700 rev/min for 15·5 hours. The cells were irradiated at 0°C with (a) 30, (b) 60 and (c) 100 krads. (d)–(f) are the associated plots of C/MΔM against *M*. Fraction 1 is at the bottom of the centrifuge tube. (Taken from Lehmann and Ormerod, 1970b.)

1. by centrifuging the gradients at 10^4 rev/min for 1 hour; under these conditions after low radiation doses (2–10 krads), the complex can be recovered on the gradient as a single sharp peak;

2. by forming the gradient above a shelf of 60% w/v sucrose saturated with CsCl and spinning at 2×10^4 rev/min for 1 hour; the complex is held on the dense shelf while the free DNA remains on the top of the gradient (see Fig. 3); or

3. by forming a gradient of CsCl-sucrose above a shelf of 60% sucrose/CsCl (density = 1·84) and spinning for 20 hours at 4×10^4 rev/min; the free DNA sediments on to the dense shelf while the complex bands isopycnically higher up the gradient at a density of about 1·4 (Fig. 4).

If the cells are labelled with ^{14}C-acetate as well as ^{3}H-thymidine, 5–10% of the incorporated ^{14}C is associated with the DNA complex (95% of the

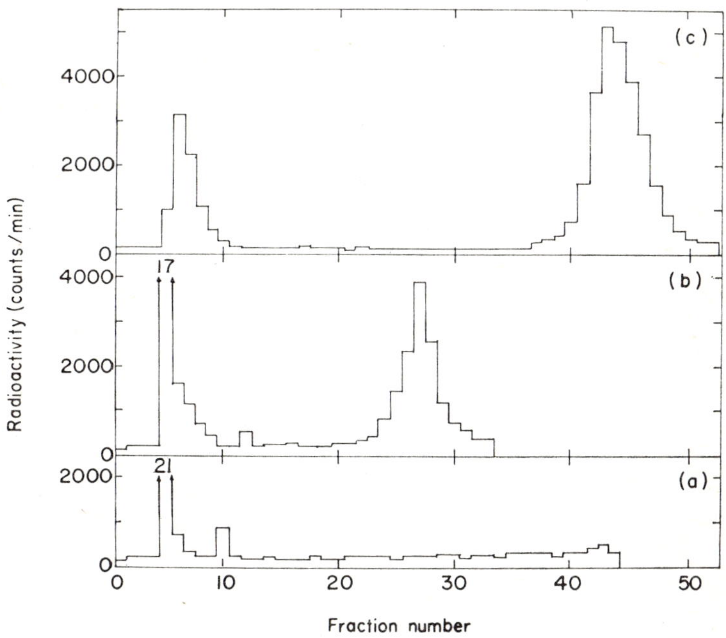

FIG. 3. Double-stranded DNA—the rapidly sedimenting DNA complex held on a CsCl-sucrose shelf. Gradient: 4·4 ml of 5–20% (w/v) sucrose, 2 M NaCl, 0·01 M sodium citrate (pH 9) formed on a shelf of 60% (w/v) sucrose saturated with CsCl. Lysing solution: as in Fig. 2. Labelled L5178Y cells. Gradients were spun at 20 000 rev/min for 1 hour. The numbers on the diagram represent 10^{-3} radioactivity found on the shelves. The cells were irradiated with X-rays; (a) 0, (b) 20 and (c) 50 krads. Note that the radiation (which introduces double-strand breaks) releases the DNA from rapidly sedimenting complex. Fraction 1 is at the bottom of the centrifuge tube. (Taken from Ormerod and Lehmann, 1971.)

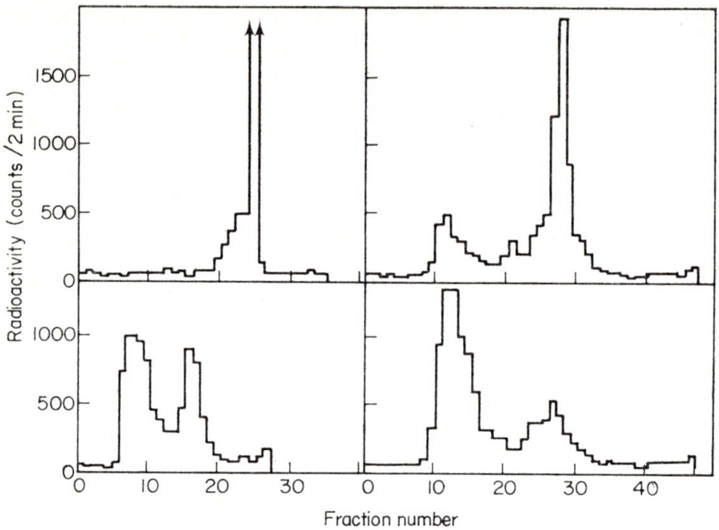

FIG. 4. Isopycnic banding of the DNA complex. Initial gradient: 0·8 ml of 60% (w/v) sucrose saturated with CsCl (density 1·84), 2·0 ml of 60% (w/v) sucrose, CsCl (density 1·56), 2·1 ml of 5–20% (w/v) sucrose, 2 M NaCl, 0·01 M sodium citrate. Lysing solution: as in Fig. 2. Labelled L5178Y cells. Gradients were spun at 40 000 rev/min for 20 hours. The cells were irradiated with X-rays (a) 0, (b) 30, (c) 100 and (d) 250 krads. The DNA bands about one-third of distance down the tube at a density of about 1·4. The free DNA sediments further down the tube to a position corresponding to a density of 1·7. Fraction 1 is at the bottom of the centrifuge tube. (Taken from Ormerod and Lehmann, 1971).

^{14}C is incorporated into lipid and less than 1% into DNA). It is the association with lipid that presumably gives the complex a buoyant density lower than that of free DNA (1·4 as opposed to 1·7).

The DNA can be released from the complex by irradiating the cells, thereby lowering the molecular weight. The rate of release as measured on a sucrose gradient (method 2 above) is more rapid than that measured on a CsCl/sucrose gradient (method 3). In order to explain this discrepancy, it has been suggested that large pieces of membrane attached to DNA cause it to sediment rapidly in sucrose while smaller pieces, which are sufficient to lower the density of the DNA, do not have the same effect on the sedimentation rate (Ormerod and Lehmann, 1971a). The picture is also complicated by the unknown extent to which high molecular weight DNA entangles with the complex.

Since radiation randomly introduces a known number of breaks into the DNA, from the measured rate of release it is possible to calculate the minimum distance between the points of attachment of the DNA to the membrane.

For L5178Y lymphoma cells this was 2×10^9 daltons (Ormerod and Lehmann, 1971a).

From the results discussed above, it is obvious that the sedimentation rate of the DNA complex cannot be related to the molecular weight of the DNA within it. Similarly, attempts to measure the efficiency of break production by radiation by measuring changes in the sedimentation rates of these complexes are meaningless. It should be noted, that the largest piece of free DNA that can be obtained by these methods has a molecular weight of 10^9 daltons.

V. Experimental Studies of Denatured DNA

A. THE RELEASE OF DENATURED DNA FROM A MAMMALIAN CELL

On alkaline sucrose gradients, the DNA from untreated cells sediments rapidly in a complex (Lett *et al.*, 1967). We can find no evidence that this complex contains other cellular components (such as lipid) and its density in alkaline caesium chloride is identical to that of free DNA (Lehmann, 1970). The rapid sedimentation is probably the result of excessive entanglement of long chains of DNA. The incomplete unwinding and separation of long sister strands is also important (McBurney and Whitmore, 1972).

In order to obtain freely sedimenting DNA the molecular weights must be lowered. This can be achieved by

a. allowing alkaline hydrolysis on top of the gradient (Lett *et al.*, 1970; Elkind and Kampfer, 1970);
b. exposing the top of an alkaline gradient to light (Elkind, 1971);
c. high energy irradiation of the cells prior to lysis;
d. treating the cells prior to lysis with a drug that introduces single-strand breaks, or
e. selectively degrading one of the two strands of the DNA molecule (McBurney and Whitmore, 1972).

The latter method which has been developed by McBurney and Whitmore (1972) prevents incomplete separation of the two strands. Mouse L cells were cultured for several generations in the presence of ^{14}C-thymidine. The cells were then grown for one generation in medium containing ^{3}H-bromodeoxyuridine (BUdR) which labelled one strand of the DNA molecule. Before layering on to an alkaline sucrose gradient, the cells were exposed to fluorescent light in order to introduce alkali-labile bonds into the Bu-substituted strand (Hutchinson and Hales, 1970; Regan *et al.*, 1971). This technique successfully separated the low molecular weight ^{3}H-labelled molecules. The ^{14}C sedimentation profile consisted of a narrow spike with a sedimentation coefficient of about 320s. The authors concluded that the shape of the profile was probably due to a sedimentation anomaly (see below) and that the DNA had a molecular weight greater than 2×10^9 daltons. This is a minimum estimate since fluorescent light introduces breaks into normal DNA (albeit with low efficiency). There is no evidence to contradict the speculation that the chromosome contains just one molecule of DNA (McBurney and Whitmore, 1972; Lett *et al.*, 1967).

In order to study the release of DNA from the fast-sedimenting complex by methods a to d above, the sucrose gradient can be formed over a shelf consisting of a 60% sucrose solution saturated with CsCl. After centrifugation (typically at 25 000 rev/min for 1 hour) the complex is found on the shelf which is denser than DNA. The free DNA will sediment in the gradient. Using this technique, we have found that 1 krad of X-irradiation to L5178Y cells releases all the DNA from the complex. This radiation dose corresponds to one break per 4×10^8 daltons.

B. ANOMALOUS SEDIMENTATION OF HIGH MOLECULAR WEIGHT DENATURED DNA

The use of high energy radiation to introduce single-strand breaks into the DNA of cells offers the most convenient method of studying molecules of high molecular weight. Radiation introduces breaks at random so that the resulting molecular weight distribution is random (Lehmann and Ormerod, 1970a). Since the number of breaks introduced is proportional to radiation dose, when the break efficiency has been measured, DNA molecules of a given average molecular weight and of known molecular weight distribution can be obtained by selecting the appropriate dose of radiation. For reasons given below, in the first instance, the break efficiency should be measured at radiation doses of 5 krad and above.

After a radiation dose of 1 krad, the DNA which has been released from the complex sediments at rotor speeds above 2×10^4 rev/min to give a sharp profile. The shape of this profile is similar to that obtained from a monodisperse collection of molecules of weight about 3×10^8 daltons (Ormerod and Lehmann, 1971; McBurney *et al.*, 1971). The same result is obtained if alkaline hydrolysis is used to release the DNA from the complex (Elkind and Kampfer, 1970; Lett *et al.*, 1970).

If a rotor speed of less than 10^4 rev/min is used, a broader profile with a higher sedimentation coefficient is obtained (McBurney *et al.*, 1971; Elkind, 1971; Lehmann and Ormerod, 1971b) (see Fig. 5). If the molecular weight of the DNA is reduced to 10^8 daltons or less (for example, by a radiation dose greater than 5 krads), the sedimentation coefficient and shape of the profile is independent of rotor speed. This effect is similar to that obtained with native DNA (see above) and is observed in a similar range of molecular weights. In view of the anomalous sedimentation of high molecular weight DNA, the sharp sedimentation profiles observed at high rotor speeds should

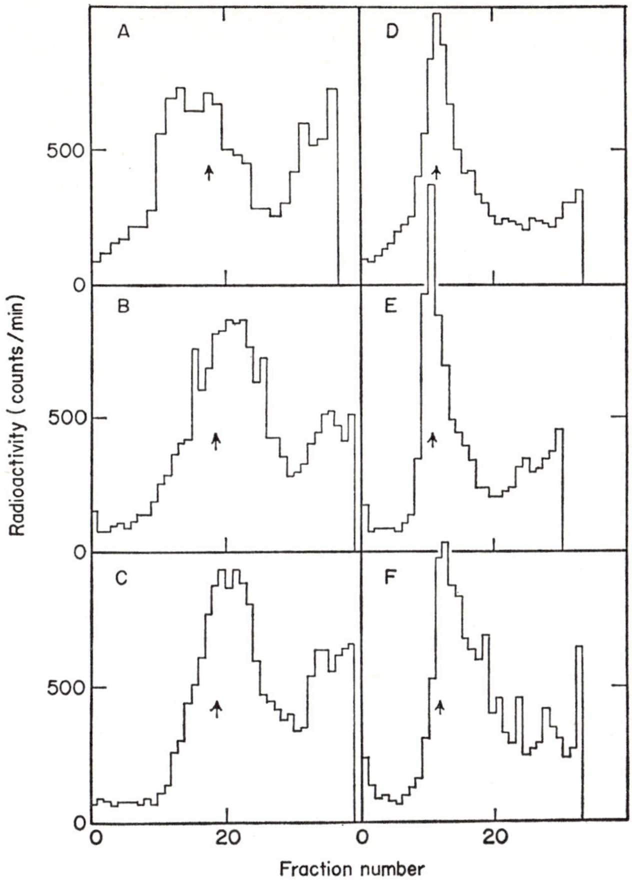

FIG. 5. Alakaline sucrose gradient profiles of ^3H-labelled DNA from irradiated L5178Y cells. Gradient: 5–20% sucrose (w/v), 0·9 M NaCl, 0·1 M NaOH. Lysing solution: 0·1% Nonidet P40, 0·1 M NaOH, 0·02 M EDTA. Gradients A–C were spun at 16 000 rev/min for 3·5 hours and D–F at 34 000 rev/min for 1 hour. A, D received 1 krad of X-rays; B, E: 2 krads and C, F: 3 krads. The arrows mark a position on the gradient corresponding to a DNA molecular weight of 3×10^8 daltons. This figure demonstrates the effect of rotor speed on the sedimentation of DNA of high molecular weight. Fraction 1 is at the bottom of the centrifuge tube. (Taken from Omerod and Lehmann, 1971b.)

not be used as evidence for the existence of sub-units of chromosomal DNA. The profiles obtained after X-irradiation and alkaline hydrolysis are identical and there is no definitive evidence to support the contention (Humphrey *et al.*, 1968; Lett *et al.*, 1970) that chromosomal DNA contains a small number of alkali-labile bonds.

C. THE EFFICIENCY OF SINGLE-STRAND BREAK PRODUCTION IN CELLULAR DNA BY HIGH ENERGY RADIATION

The number of single-strand breaks introduced into DNA by high energy radiation should be estimated from a linear plot of $1/M_n$ against radiation dose. The lowest radiation dose used should be high enough (probably 5 krads) to ensure that the sedimentation rate of the DNA is independent of rotor speed and the cells should be irradiated in ice to prevent any rejoining of strand breaks during irradiation. Table I shows some of the results that have been obtained with different cells. Any data that can be objected to on the grounds given here or in the Analysis section have been ignored. It can be seen that on average one strand break is introduced for every 60 eV of absorbed radiation.

TABLE I

The efficiency of production of single-strand breaks in the DNA of X-irradiated mammalian cells

Cell	eV per break	Reference
Chinese hamster ovary	60	Lett and Sun (1970)
Murine lymphoma (L5178Y)	44	Lehmann and Ormerod (1970a)
Human lymphocytes	55	Donlon and Norman (1971)
Murine L cells	72	McBurney *et al.* (1971)

Gassing with either oxygen or nitrogen has no effect on the number of strand breaks introduced into the DNA of L5178Y cells (Ormerod and Stevens, 1971). A careful study of synchronised Chinese hamster ovary cells by Lett and Sun (1970) showed that the efficiency of strand breakage was the same throughout the cell cycle.

Sowada and Okada (1970b) found that cysteamine at a concentration greater than 5 mM protected the DNA of L5178Y cells irradiated at 37°C. At a concentration of 50 mM they obtained a dose reduction factor of six. Cystamine had no effect. Lohman *et al.* (1970) obtained similar results with human T cells irradiated at room temperature (both cysteamine and glycerol gave maximum dose reduction factors of four). In contradiction to these results, we obtained no effect on L5178Y cells with either cysteamine or cystamine (Ormerod and Stevens, 1971). We irradiated at 0°C and the discrepancy in the results might be due to the different temperatures of irradiation. In view of the importance of cysteamine as a radio-protective agent, its effect on strand breakage needs further investigation.

D. THE REJOINING OF SINGLE-STRAND BREAKS

The rapid rejoining of single-strand breaks in the DNA of irradiated cells has been reported by many authors working with a variety of cell lines in tissue culture (Lett *et al.*, 1967; Lohman, 1968; Humphrey *et al.*, 1968; Matsudaira *et al.*, 1969; Sowada and Okada, 1970; Kleijer *et al.*, 1970; Ormerod and Stevens, 1971; Elkind, 1971; Donlon and Norman, 1971; Horikawa *et al.*, 1972). The results are entirely consistent with the hypothesis that all cells capable of division can rejoin radiation-produced strand breaks. Fifty per cent of the strand breaks are rejoined in about 10 to 20 min and within 1 hour at least 90% of the breaks are repaired (Ormerod and Stevens, 1971; Donlon and Norman, 1971; Horikawa *et al.*, 1972).

The rate of rejoining is temperature-dependent. The temperature has to be lowered to 5°C or below to completely inhibit the reaction (Sowada and Okada, 1970; Ormerod and Stevens, 1971). Donlon and Norman (1971) measured an activation energy of $19 \pm 0{\cdot}4$ krad/mole for the initial rejoining reaction in the temperature range 22–37°C in human lymphocytes. They observed a maximum rate of 40°C.

Rejoining of strand breaks is not affected by inhibition of protein, RNA or DNA synthesis. Inhibitors used included hydroxyurea (Sowada and Okada, 1970), thymidine (Sowada and Okada, 1970; Ormerod and Stevens, 1971), mitomycin C (Sowada and Okada, 1970), actinomycin D (Sowada and Okada, 1970; Ormerod and Stevens, 1971; Elkind, 1971) and puromycin (Sowada and Okada, 1970; Ormerod and Stevens, 1971). 2,4-dinitrophenol has no effect on strand-break repair in L5178Y cells which have been given a radiation dose of 10 krads (Sowada and Okada, 1970; Ormerod and Stevens, 1971), but after 30 krads it blocks the rejoining of about 50% of the breaks (Ormerod and Stevens, 1971). This suggests that ATP is necessary for the rejoining process and that, prior to irradiation, the nucleus contains enough ATP to rejoin the number of breaks introduced by 15 krads.

The results indicate that the enzymes responsible for repairing radiation-produced single-strand breaks are not induced but are always present within the cell. It is possible that these enzymes are involved in the normal replication of DNA. Alternatively, they might be part of a system which constantly monitors DNA and repairs defects as they arise. This would help maintain both the genetics stability of a species and the information content of a cell during the life of an animal.

We have found that the peripheral blood lymphocytes of a chicken (the majority of which are not in cell cycle but are capable of being stimulated to divide) rejoin single-strand breaks. The chicken erythrocytes (which are post-mitotic) cannot repair strand breaks (Karran and Ormerod, 1972). These results lend support to the hypothesis that only fully differentiated cells are incapable of strand break repair.

VI. The Introduction of Strand Breaks by Agencies other than High Energy Radiation

Alkaline sucrose gradients have been used to observe the breakdown of DNA in mammalian cells caused by u.v. light (Setlow *et al.*, 1969; Cleaver and Thomas, 1969; Ben-Hur and Ben-Ishai, 1971; Regan *et al.*, 1971), N-ethyl maleimide, methyl hydrazine (Moroson and Furlan, 1969a, b), methyl methane sulphonate (Strauss *et al.*, 1969; Fox and Fox, 1970), and 4-hydroxyl-amino quinoline-l-oxide (Horikawa *et al.*, 1970, 1972). Most of the foregoing comments are applicable to these studies and this section is restricted to a few observations of particular interest.

Ultraviolet directly introduces few strand breaks into DNA; the major products are pyrimidine dimers which are excised from the DNA in human cells (for example, see Regan *et al.*, 1968). The initial step in the excision process, the insertion of a break by specific endonuclease, can be demonstrated by sucrose gradient sedimentation (Setlow *et al.*, 1969; Ben-Hur and Ben-Ishai, 1971). After the incision, the dimer is excised, the resulting gap filled by new synthesis and the original break is rejoined (e.g. see Setlow, 1968). The latter step results in a return of the DNA to its original molecular weight.

The rates of the different enzymic reactions involved appear to show considerable variation between different cell lines. Setlow *et al.* (1969) observed a maximal amount of DNA degradation in u.v. irradiated skin fibroblasts after about 10 hours; rejoining was completed after 24 hours. In Hela cells, the corresponding times were 90 min and 8 hours (Ben-Hur, 1971) while Kleijer *et al.* (1970) observed no strand breaks in the DNA of u.v. irradiated human kidney (T) cells.

In some cell lines, notably those from rodents, the excision–repair process is inefficient. In these cells the DNA synthesised after u.v. irradiation is of lower molecular weight compared to that from unirradiated cells (Cleaver and Thomas, 1969; Lehmann, 1972) suggesting that the newly synthesised DNA contains gaps. Subsequently these gaps are filled in. This effect has been demonstrated by sucrose gradient analysis but a quantitative analysis is complicated (Lehmann, 1972) because the apparent molecular weight distribution from end-labelled differs from that of fully labelled molecules (Lehmann and Ormerod, 1969).

Another method for studying the effects of u.v. on DNA, which involves sucrose gradient analysis, utilises the observation that u.v. light at 313 nm wavelength preferentially introduces single-strand breaks into BUdR-labelled DNA (Hutchinson and Hales, 1970). Regan *et al.* (1971) irradiated skin fibroblasts at 254 nm. The cells were then incubated in the presence of hydroxyurea and BUdR for 24 hours during which time the repaired sections of DNA are patched with short lengths of nucleotides including BUdR.

Subsequent irradiation at 313 nm introduced strand breaks at the repaired sites. This method is particularly suitable for quantitatively comparing the efficiency of repair replication in different types of cells.

Radiation (either high energy or u.v.) allows the delivery of a given dose to cells at 0°C. Subsequent enzymatic reactions can be followed at a higher temperature. If the effects of cytotoxic drugs (for example, methyl methane sulphonate) are being studied, the drug treatment is usually only effective at temperatures high enough to permit metabolic activity. Consequently, chemical damage, nuclease attack and DNA repair can all occur during the time of treatment. This complicates quantitative measurement and makes it difficult to unequivocally interpret the results.

Despite these limitations, it is possible to gain useful information from a sucrose gradient analysis. An example of this is shown in Fig. 6 in the case of degradation of DNA in L5178Y cells induced by heating at 43°C. If breaks are introduced at random into the DNA of all the cells in a population, the

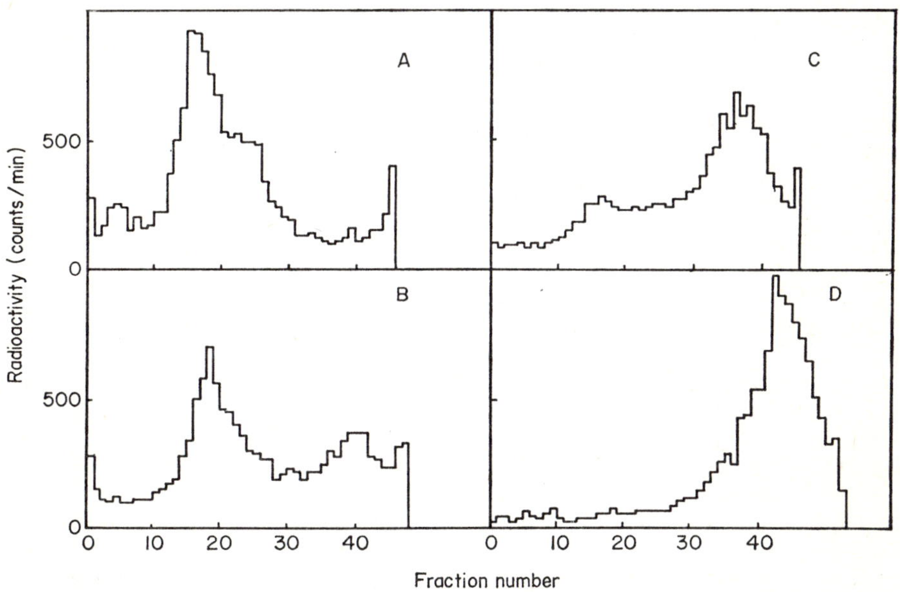

FIG. 6. The effect of incubation at 43°C on the DNA from irradiated L5178Y cells. Gradient and lysing solution: as in Fig. 1. Gradients spun at 37 000 rev/min for 1 hour. Radiation dose (X-rays): 10 krads. Incubated after irradiation in growth medium at 43°C for A: 20, B: 40, C: 60 and D: 90 min. In the first 20 min after irradiation, 70% of the single-strand breaks rejoined to give a M_n of about 10^8 daltons. Subsequently, the DNA degraded. The DNA disappeared from the main peak and re-appeared on top of the gradient. Fraction 1 is at the bottom of the gradient. (Taken from Ormerod and Stevens, 1971.)

resulting molecular weight distribution will be random and as the time of treatment is extended, the peak of the DNA profile will sediment more slowly. Radiation gives a typical example of this behaviour. Heat treatment had a totally different effect. The main DNA peak did not alter in molecular weight; DNA was lost from it and appeared on the top of the gradient as short acid-soluble fragments. The results indicated that some random event occurred in a cell which caused a rapid breakdown of its DNA. The random event was not occurring at the level of the DNA molecule but at a cellular level. (The data in Fig. 6 are from irradiated cells; similar results can be obtained from unirradiated cells; G. M. Jolley, unpublished work.) This difference in the modes of degradation of DNA is clearly demonstrated by sucrose gradient sedimentation.

Acknowledgments

This work was supported by grants to the Chester Beatty Research Institute (Institute of Cancer Research: Royal Cancer Hospital) by the Medical Research Council and the Cancer Research Campaign.

References

Abelson, J. and Thomas, C. A. jr. (1966). *J. molec. Biol.* **18**, 262–288.

Ben-Hur, E. and Ben-Ishai, R. (1971). *Photochem. Photobiol.* **13**, 337–345.

Britten, R. J. and Roberts, R. B. (1960). *Science, N.Y.* **131**, 32–33.

Burgi, E. and Hershey, A. D. (1963). *Biophys. J.* **3**, 309–321.

Charlesby, A. (1954). *Proc. R. Soc. A* **224**, 120–128.

Cleaver, J. E. and Thomas, G. H. (1969). *Biochem. biophys. Res. Commun.* **36**, 203–208.

Dean, C. J., Omerod, M. G., Serianni, R. W. and Alexander, P. (1969). *Nature, Lond.* **222**, 1042–1044.

Donlon, T. and Norman, A. (1971). *Mutation Res.* **13**, 97–107.

Doty, P., McGill, B. B. and Rice, S. A. (1958). *Proc. natn. Acad. Sci. Wash.* **44**, 432–438.

Elkind, M. M. (1971). *Biophys. J.* **11**, 502–520.

Elkind, M. M. and Kampfer, C. (1970). *Biophys. J.* **10**, 237–245.

Fox, B. W. and Fox, M. (1969). *Mutation Res.* **8**, 629–638.

Horikawa, M., Nikaido, O., Tanaka, T., Nagata, H. and Sugahara, T. (1970). *Expl Cell Res.* **59**, 147–152.

Horikawa, M., Fokuhara, M., Suzuki, F., Nikaido, O. and Sugahara, T. (1972). *Expl Cell Res.* **70**, 349–359.

Humphrey, R. M., Steward, D. L. and Sedita, B. A. (1968). *Mutation Res.* **6**, 459–465.

Hutchinson, F. and Hales, H. B. (1970). *J. molec. Biol.* **50**, 59–69.

Karran, P. and Ormerod, M. G. (1973). *Biochim. biophys. Acta* **299** 54–64.

Kleijer, W. J., Lohman, P. H. M., Mulder, M. P. and Bootsma, D. (1970). *Mutation Res.* **9**, 517–523.

Lehmann, A. R. (1970). Ph.D. thesis, University of London.

Lehmann, A. R. (1972). *J. molec. Biol.* **66**, 319–337.

Lehmann, A. R. and Ormerod, M. G. (1969). *Nature, Lond.* **221,** 1053–1056.
Lehmann, A. R. and Ormerod, M. G. (1970a). *Biochim. biophys. Acta* **204**, 128–143.
Lehmann, A. R. and Ormerod, M. G. (1970b). *Biochim. biophys. Acta* **217**, 268–277.
Leighton, S. B. and Rubenstein, I. (1969). *J. molec. Biol.* **46**, 313–328.
Lett, J. T., Caldwell, I. W. R., Dean, C. J. and Alexander, P. (1967). *Nature, Lond.* **214**, 790–792.
Lett, J. T., Klucis, E. S. and Sun, C. (1970). *Biophys. J.* **10**, 277–292.
Lett, J. T. and Sun, C. (1970). *Radial. Res.* **44**, 771–787.
Lohman, P. H. M., Vos, O., Van Sluis, C. A. and Cohen, J. A. (1970). *Biochim. biophys. Acta* **224**, 339–352.
Lohman, P. H. M. (1968). *Mutation Res.* **6**, 449–458.
McBurney, M. W., Graham, F. L. and Whitmore, G. F. (1971). *Biochem. biophys. Res. Commun.* **44**, 171–177.
McBurney, M. W. and Whitmore, G. F. (1972). *Biochem. biophys. Res. Commun.* **46**, 898–904.
McGrath, R. A. and Williams, R. W. (1966). *Nature, Lond.* **212**, 534–535.
Matsudaira, H., Nakagawa, C. and Bannai, S. (1969a). *Int. J. Radiat. Biol.* **15**, 575–581.
Matsudaira, H., Nakagawa, C. and Hishiwaza, T. (1969b). *Int. J. Radiat. Biol.* **15**, 95–100.
Moroson, H. and Furlan, M. (1969a). *Radiat. Res.* **40**, 351–365.
Moroson, H. and Furlan, M. (1969b). *Physiol. chem. Physics.* **1**, 368–376.
Moroson, H. and Furlan, M. (1970). *Radiat. Res.* **44**, 713–726.
Noll, H. (1967), *Nature, Lond.* **215**, 360–363.
Ormerod, M. G. and Lehmann, A. R. (1971b). *Biochim. biophys. Acta* **247**, 369–372.
Omerod, M. G. and Lehmann, A. R. (1971a). *Biochim. biophys. Acta* **228**, 331–343.
Ormerod, M. G. and Stevens, U. (1971). *Biochim. biophys. Acta* **232**, 72–82.
Regan, J. D., Setlow, R. B. and Ley, R. D. (1971). *Proc. natn. Acad. Sci. U.S.A.* **68**, 708–712.
Sawada, S. and Okada, S. (1970b). *Radiat. Res.* **44**, 116–132.
Sawada, S. and Okada, S. (1970a). *Radiat. Res.* **41**, 145–162.
Setlow, R. B. (1968). *Prog. Nucl. Acid. Res. molec. Biol.* **8**, 257–295.
Setlow, R. B., Regan, J. D., German, J. and Carrier, W. L. (1969). *Proc. natn. Acad. Sci. U.S.A.* **64**, 1035–1041.
Strauss, B., Coyle, M., and Robbins, M. (1968). *Cold. Spring Harb. Symp. quant Biol.* **33**, 277–287.
Studier, F. W. (1965). *J. molec. Biol.* **11**, 373–390.
Terasima, T. and Tsuboi, A. (1969). *Biochem. biophys. Acta* **174**, 309–314.
Veatch, W. and Okada, S. (1969). *Biophys. J.* **9**, 330–346.
White, H. J. (1971). Ph.D. thesis, University of London.

Note added in proof

Since this review was completed, two groups of workers†‡ have reported an oxygen effect on the production of single-strand breaks in the DNA of mammalian cells irradiated with X- or γ-rays. The cells used were murine fibroblasts (L-60)† and chinese hamster (Line VA-379A).‡

† Palcic, B. C. and Skarsgard, L. D. (1972). *Int. J. Rad. Biol.* **21**, 417–433.

‡ Dugle, D. L., Chapman, J. D., Gillespie, L. J., Borsa, J., Webb, R. G., Meeker, B. E. and Reuvers, A. P. (1972). *Int. J. Rad. Biol.* **22,** 545–555.

CHAPTER 22

Stability and Evolution of DNA from the point of view of Molecular Radiobiology

M. RADMAN, J. ROMMELAERE† AND M. ERRERA

Laboratoire de Biophysique et Radiobiologie, Faculté des Sciences, Université libre de Bruxelles

† Aspirant du Fonds National de la Recherche Scientifique.

Several remarkable reviews on DNA repair and mutagenesis have been written in the past years covering all important research in this field (Howard-Flanders and Boyce, 1966; Setlow, 1968; Howard-Flanders, 1968; Witkin, 1969; Drake, 1970; Smith, 1971; Smith and Hanawalt, 1969). Since no major discoveries were made in the meantime, we will discuss in a very general context only the data which we consider the least ambiguous. We shall neither attempt to give a critical evaluation of all the published data. This was done by the above cited authors. These restirctions will necessarily lead to omit some promising work and ideas.

The main strategy we wish to pursue is to present some basic facts and models from recent molecular radiobiological studies on DNA repair and mutagenesis in relation to general evolutionary problems. Recent books (Fox, 1965; Calvin, 1969; Kimbal and Oro, 1971) and reviews (Orgel, 1968; Crick, 1968; Eigen, 1971a, b; Spiegalman, 1971) on molecular aspects of evolution have not considered that DNA repair may have played an important role. We shall eventually suggest some working hypotheses for sake of the discussion and as an arbitrary way to rationalise our present knowledge.

I. Some General Evolutionary Problems: Outline of the Present Review

The evolution of nucleic acids as carriers of genetic information was presumably directly dependent on the parallel evolution of mechanisms for their protection and repair. In fact, several features of repair processes such as the molecular basis of their action, the complexity of the molecules involved, their phylogenetic distribution, independently suggest an evolutionary pathway from very simple protection or repair to highly sophisticated enzymatic error correction processes: any knowledge of these pathways might help to understand at least some stages of the evolution of the genetical material.

At the very early stages of the chemical prebiological evolution, the main problem was to accumulate and preserve molecules capable of carrying information rather than finding a way for this information to vary. We will briefly review some essential work on the simulation of prebiotic synthesis which suggests the importance of solar radiation energy in the formation of the first elements of nucleic acids, and the possible role that other organic molecules, also formed in abiotic conditions, may have played in their maintenance.

At later stages, variability and selection by functional competition must have greatly determined the rate of evolution. Protection of evolved molecular structures against their destruction or too extensive change, required the emergence of more sophisticated repair mechanisms. Error-proof repair processes are the requirements for stabilising genetic information whilst error-prone repair mechanisms provide the elements for its variability; this will be discussed in the light of recent hypotheses on the molecular mechanisms of mutagenesis and of genetic recombination. We will also discuss the necessary coordination between the essential DNA functions such as replication, recombination, repair (and transcription); any failure of the coordination of these processes on a given segment of DNA is expected to be fatal.

Finally, as organisms became more and more complex, the storage and maintenance of the genetic material, more vulnerable on account of its increasing size, became still more critical and an attempt will be made to outline the role its distribution into chromosomes and the appearance of sexuality may have played in maintaining this information intact or in promoting its variability.

The text is organised according to the phenomena and processes relevant to DNA repair and mutagenesis rather than according to the organisms themselves; most of the studies reported were carried out with bacteria and bacteriophage but there is growing evidence that these phenomena are of universal importance since many have been discovered in yeasts and mammalian cells as well.

II. Prebiotic Evolution

A. ABIOTIC SYNTHESIS OF SIMPLE MOLECULES OF BIOLOGICAL INTEREST

1. *Plausible Energy Sources*

Miller and Urey (1959) and Hulett (1969) presented a table (Table I) which shows the spectrum of energies which probably were available for prebiotic

TABLE I

Present sources of energy averaged over the earth. (From Miller and Urey, 1959.) *Assuming that original organic matter was synthesised from atmospheric CH_4, NH_3, H_2O, CO_2, etc., the main energy source was probably sunlight <200 nm. When more complex molecules absorbing longer wavelength were formed, u.v. (~ 250 nm) could have driven major synthetic photochemical reaction which eventually led to photosynthesis*

Source	Energy (cal cm^{-2} yr^{-1})
Total radiation from sun	260,000
Ultraviolet light	
$\lambda < 250$ nm	570
$\lambda < 200$ nm	85
$\lambda < 150$ nm	3·5
Electric discharges	4
Cosmic rays	0·0015
Radioactivity (to 1·0 km depth)	0·8
Volcanoes	0·13

syntheses. It is striking that u.v. light was probably present in great excess over all non-electromagnetic sources of energy; thus u.v. can be considered as one of the most likely energy source for prebiological synthesis. However, the importance of other sources (electrical discharges, radioactivity, heat, etc. . . .) cannot be overlooked. Sagan (1965) calculated that about 4×10^9 years ago, the reducing atmosphere was likely to have left a "window" for u.v. at wavelengths 240–290 nm (u.v. being cut off above 290 by formaldehyde). This "window" was closed much later, probably as a consequence of biological evolution (assimilation by green plants), by the formation of an ozone layer in the upper atmosphere.

2. *Ultraviolet Light as Energy Source*

The suspected primary constituents of biological significance in the primitive reducing atmosphere are hydrogen, methane, ethane, ammonia, water and possibly carbon monoxide and nitrogen. By irradiation of these constituents with short u.v. light, primarily cyanides, aldehydes and hydrocarbons can be produced. Purines as well as ribose and deoxyribose can be obtained by further u.v. or γ irradiation or by electron bombardment (Ponnamperuma, 1965). Purines, uracil as well as cytosine and a variety of sugars can also be obtained chemically from simple precursor molecules at moderate temperatures (Oro, 1965; Sanchez *et al.*, 1966; Ferris *et al.*, 1968). Thymine was synthesised more recently, in possible prebiotic conditions, as a result of uracil methylation (Stephen-Sherwood *et al.*, 1971). Under these conditions, adenine is synthesized more easily and in larger quantities than other bases. Of all nucleic acid bases adenine has the greatest resonance energy (Pullman and Pullman, 1962) and therefore is the most radiation resistant. This may be the reason for the large distribution and important roles of adenine derivatives (ATP, cAMP, . . .) in living cells (Ponnamperuma *et al.*, 1963).

Ultraviolet irradiation of a solution of adenine and ribose acid produces adenosine, and this, in the presence of ethyl metaphosphate, yields AMP, ADP and ATP (Sagan, 1965; Ponnamperuma, 1965). Phosphorylation of nucleosides by moderate heating was also obtained in the presence of microcrystals of silicates or apatite, acting perhaps as primitive catalysts (Neuman *et al.*, 1970), and a mixture of adenosine monophosphates was produced by exposing adenosine to the recoiled ^{31}P atoms from ^{31}Si decay (Akaboshi *et al.*, 1971). More recently still, cytidine nucleotides and oligonucleotides (Tapiero and Nagyavary, 1971), thymidine nucleotides (Bishop *et al.*, 1972) as well as glucose 1-6-diphosphate were also obtained in possible prebiotic conditions; however phosphorylation of nucleotides in view of further polymerisation remains one of the key problems and is under intensive study (Rabinowitz *et al.*, 1971; Schwartz and Ponnamperuma, 1971).

Using u.v. (253·7 nm) as the energy source and hydrogen sulphide or formaldehyde as photon acceptor, Steinman *et al.* (1968) and Sagan and Khare (1971) obtained a highly efficient photoproduction of amino acids. On the other hand, in classical electric discharge experiments (Miller, 1955), amino acids were synthesized but no u.v. absorbing material similar to heterocyclic bases was obtained. Heat could also have played an important role, as in the condensation of glycerol with a hydrogen cyanide tetramer to form a dipeptide (Chang *et al.*, 1969) or more complex peptides (Fox, 1964).

3. *Polymerisation*

Starting with mononucleotides, polymerisation to oligo- and polynucleotides can be achieved by various, possibly prebiotic, agents such as u.v. irradiation (McReynolds *et al.*, 1971) or cyanamide (Ibanez *et al.*, 1971). Stacked base interactions (Bugg *et al.*, 1971), crystalline catalysts like montmorillonite (Ibanez *et al.*, 1971) or genuine templates (like polycytidylate or polyuridilic acid) (Schneider-Bernloehr *et al.*, 1968; Sulston *et al.*, 1969) might all have contributed to the formation of oligo or polynucleotides. The association of complementary polynucleotides could have led to the formation of longer duplexes (Calvin, 1969), either strand of which could in turn have served as template for the organisation of two new duplexes.

Katchalsky and his group synthesised polyalanine from alanyladenylate on montmorillonite and the role of these crystals as a primitive template is suggested by the fact that one obtains essentially nonamers or chains of multiple length (Paecht-Horowitz *et al.*, 1970). Simple organic polyamines or glycine derivatives may have contributed not only to the stabilisation of nucleotides on the template but to the catalytic formation of internucleotide bonds as observed with the synthesis of poly A from adenosine cyclic 2′-3′-phosphate units organised as a third helix on 2 poly U strands (Renz *et al.*, 1971).

4. *Genetic Code*

The general nature of the genetic code may have been determined very early before the appearance of protein catalysts. That "proteins function but apparently cannot replicate, that nucleic acids replicate but apparently cannot function" led Orgel and Sulston (1971) to postulate a connection between polynucleotides and polypeptides from the very beginning.

The specific recognition problem related to the nature of the genetic code was recently discussed in a very broad context by Eigen (1971). Findings from his laboratory (Pörschke and Eigen, 1971) indicate that three A–U pairs or two G–C pairs represent the minimum recognition length necessary for nucleation in the process of annealing single strands of nucleic acids. According to Eigen (1971), a genetic code made of sequences smaller than a triplet would be restricted by the physical nature of oligonucleotide recognition; on the other hand large single stranded recognition units would have the tendency to collapse into partially coiled structures and in addition, matching of long coding units with the "messenger" strand would lead to the formation of rather stable duplexes. Apparently, a triplet code can satisfy both the recognition properties and the flexibility of the translational process, as well as the coding "logic" for amino acids. In addition, it is much more likely, on the basis of the principle of continuity, that the original coding unit was a triplet (probably much less specific than at present) because any change as drastic as that from a doublet to a triplet code would have led to the loss of all accumulated information (Crick, 1968; Orgel, 1968).

Other considerations have suggested that more or less specific photochemical addition products between amino acids and nucleotides might have played a role in the transfer of primitive information between nucleic acids and amino acids (the specificity of the code would then derive from the specificity of photochemical reactions) (Lesk, 1970). Amino acids might also interact with weak selectivity with nucleotides (Woese *et al.*, 1966; Saxinger and Ponnamperuma, 1971; Black, 1971).

B. DAMAGE TO NUCLEIC ACIDS

We will attempt to describe briefly the chief types of lesions encountered in DNA; recent reviews on radiation damage being available (Smith and Hanawalt, 1969; Kanazir, 1969; Latarjet, 1972), we will limit our references to these and only quote data they do not cover. Examples of damages inflicted by chemicals have also been briefly considered because these lead to biological effects often similar to those of radiation and because common repair mechanisms apparently exist.

1. *Damage to Bases*

The quantum yield for photochemical damage to purines is inferior by a factor 10 (for u.v.) and 2 (for ionising radiations) to that for pyrimidines. Damage to RNA is less well known, but the principal u.v. lesions found in DNA have been identified in RNA.

a. *Pyrimidine dimerisation*: by far the most frequent damage caused by u.v. light is the dimerisation of two adjacent pyrimidines by the formation of a cyclobutane ring after the breakage of the pyrimidine C_5–C_6 double bond; dimerised cytosine tends to deaminate during the acid hydrolysis necessary for their detection. Thymine dimers ($\overline{TT}$) are usually more frequently found than $\overline{CT}$ (UT), and $\overline{CC}$ (UU) occur less often ($\sim 10\%$), but this distribution of dimers depends on the primary sequence. Dimerisation is favoured at higher wavelengths ($\sim$280 nm) but dimers tend to revert to monomers at

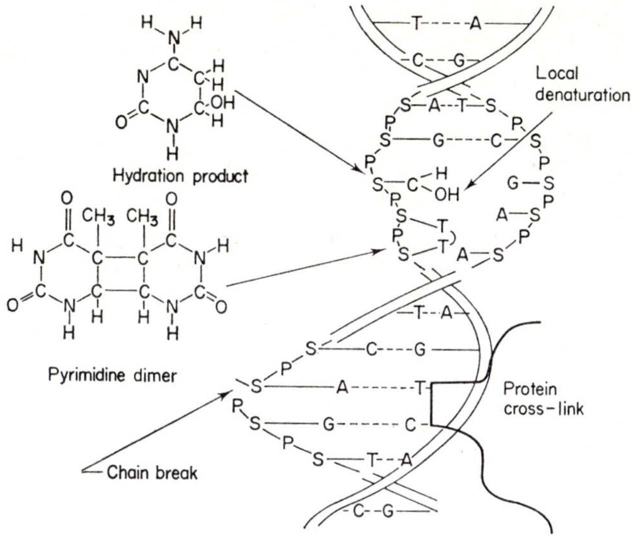

FIG. 1. Summary of u.v. damage to DNA (adapted from Deering (1962) by Smith and Hanawalt (1969)).

short wavelength (240 nm) and an equilibrium between the forward and reverse reaction is found (Deering, 1962) (see Fig. 2). Similar pyrimidine dimers can be formed by other photochemical or chemical reactions, leading to the same type of molecular excited states (Lamola, 1971). Dimerisation leads to a local distortion of the double helix. Pyrimidine dimers with unusual configuration may be obtained, as in the DNA of bacterial spores. Dimerisa-

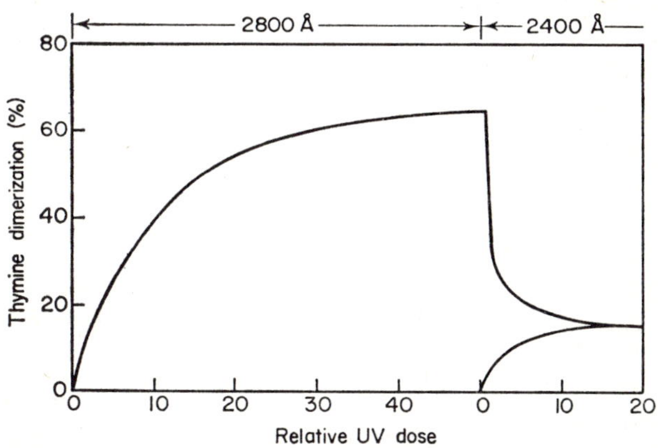

FIG. 2. Formation of thymine dimers in DNA at 280 and 240μm. The equilibrium between the formation of dimers and their monomerisation is in favour of the latter reaction at lower wavelength (from Deering, 1962).

tion may also occur between pyrimidines of both strands (as when DNA is denatured or irradiated dry, or with high intensities) leading to intra- or intermolecular crosslinking.

b. *Addition products (adducts)*: (i) Hydration of the pyrimidine C_5C_6 double bond occurs with u.v. (essentially in denatured DNA); this probably leads to mispairing. Hydrates may be lethal in viral RNA such as in tobacco mosaic virus. Other addition products are found after photochemical reaction with solutes like alcohols (R–OH) or HCN. (ii) Peroxydation of thymine (on the C_5–C_6 double bond) is the best characterised base damage caused by ionising radiation in aerated solution; hydrates as well as other saturated base derivatives can occur after oxic or anoxic irradiation besides a variety of more or less stable compounds arising after ring cleavage. (iii) Cross-linking to proteins or amino acids can be the result of a photochemical reaction between an amino acid SH or OH group and the C_5C_6 double bond of pyrimidines; proteins are frequently cross-linked to nucleic acids when cells are irradiated and this damage may be a cause of u.v. lethality in RNA viruses. (iv) Strong electrophilic reagents like alkylating agents (Lawley, 1966) efficiently attack nucleophilic centres of nucleic acids (the N_7 of guanine, and to a lesser extend N_3 of adenine) leading to the alkylation of these nitrogens; this can be followed by depurination. Bifunctional agents may give intra- (usually on the same strand) and intermolecular cross linking. (v) Hydroxylamine, also, a potent nucleophilic reagent forms adducts with pyrimidines, essentially cytosine, which may then pair with adenine (Phillips and Brown, 1967).

c. *Deamination*: HNO_2 deaminates bases like A, C and G which acquire new base pairing capacities; the formation of xanthine from guanine weakens the glycosidic bond; if hydrolysed, an aldehyde is formed capable of reacting with an amino group in the opposite chain; this may be the mechanism of interstrand cross-linking frequently observed (Burnotte and Verly, 1971).

2. *Damage to the Sugar Phosphate Backbone*

a. *Photochemical reaction in halogenated pyrimidine*: ribose and deoxyribose do not absorb u.v. above 230 nm, therefore direct photochemical reactions within the sugar in the backbone are biologically insignificant with usual u.v. sources. However, chain breakage can occur when T is replaced by 5 halouracil which shifts the absorption towards slightly longer wavelength; u.v. light even above 300 nm, can remove the halogen, yielding a uracil radical which in the absence of hydrogen donors, attacks the sugar ring with concomittant chain breakage (Hotz and Reuschl, 1967; Hutchinson and Hales, 1970).

b. *Strand breakage with ionising radiation*: several sugar radicals have been identified in deoxyribose as a result of the removal of an H atom from the sugar ring which opens at the oxygen bridge: this can be the consequence of a direct effect or be caused by radicals from irradiated water. This may also lead to chain scission sometimes correlated with biological inactivation.

c. *Strand breakage during chemical reactions*: alkylated bases are susceptible to hydrolysis as mentioned above (depurination), which could in turn lead to the hydrolysis of the deoxyribose (Lawley, 1966). The carcinogenic agent 4 hydroxylamino-1-oxyde (Sugimura *et al.*, 1968) may also be the cause of single strand breaks, which can occur under a variety of other conditions.

d. *Double-strand breakage*: double-strand breakage may occur from independent breakage of both strands opposite each other or probably within a distance of a few nucleotides; it has also been found that double-strand breakage could occur as the result of a single initial event.

3. *Local Change in Double Helical Structure*

Change in secondary and tertiary structure occurs as seen above from the presence of pyrimidine dimers which seem to occur in clusters (Shafronskaya *et al.*, 1973) or from local denaturation in the neighbourhood of single-strand breaks. Intercalation of aromatic compounds (acridines, etc.) between the planes of adjacent bases leads to doubling of the inter-nucleotidic distance (Lerman, 1963) which locally modifies the degree of torsion of the double helix (Saucier *et al.*, 1971); the various lesions found in u.v. irradiated DNA are summarized in Fig. 1 (see p. 174).

We have not attempted to give an exhaustive enumeration of damages encountered in DNA; what is certain is that different agents cause similar biological events (death, mutation, malignancy, etc.). A small number of mechanisms may be efficient in repairing a variety of primary lesions: damage caused to yeasts by nitrogen mustard (HN_2) seems to have common repair pathways with u.v. lesions whilst methylmethane sulphonate induced damage probably shares common repair pathways with ionising radiation induced lesions (Brendel *et al.*, 1970). Similarly, heating *E. coli* for a short time leads to enzymatic DNA strand breakage and rejoining, presumably by a mechanism operating for other damages as well (Woodcock and Grigg, 1972).

4. *Sunlight as Inducer of Pyrimidine Dimers*

Exposure to sunlight during 30 sec can be as lethal as 0·5 erg/mm^2 of u.v. (254 nm) to a strain of *E. coli* presumably incapable of all dark repair mechanisms; these sunlight treated cells can be extensively photoreactivated, which strongly suggests that the damage consists of pyrimidine dimers in the DNA. The corresponding wild type bacterial strain can similarly be affected by sunlight, but by doses a hundred-fold greater (Harm, 1969). A similar observation has been made in yeasts (Resnik, 1970). Pyrimidine dimers can also be formed in cultured mammalian cells (Trosko *et al.*, 1965) as well as in human skin (Pathak, 1972) in comparable conditions. This sensitivity to sunlight led Singer and Ames (1970) to suggest that natural exposure to sunlight might give sufficient selective pressure to influence the evolution of bacteria towards a high GC content of their DNA which makes them more resistant to u.v. light (Haynes, 1964); but this has recently been questioned (Leth Bak *et al.*, 1972). At present it seems difficult to make valid comparisons of survival data between species because they may have repair systems with a wide range of efficiency.

III. Elements of Stabilisation of Genetic Information

A. PREBIOTIC SYNTHESIS AND THE SURVIVAL PROBLEM

The reasons for assuming that u.v. may have been a major source of energy for the synthesis of the first prebiological organic molecules, in particular, nucleic acids, were given above. The absorption maximum of purines and pyrimidines is in the middle of the u.v. spectrum irradiating the primitive earth, thus this u.v. energy would have led to the chemical activation of these bases, i.e. to further change or to the destruction of the original structure. No generation of order and of information could have arisen without some mechanism of protection of the randomly appearing molecular structures. We have so far discussed the formation and accumulation of primitive molecules which eventually acquired biological significance. At this stage of early evolution, selection was essentially based on the chemical properties of molecules, on their association with one another and with primitive catalysts or templates and on their affinity for protecting molecules. This could have favoured the accumulation of informative sequences having the potentialities for self replication (Orgel, 1968). The selective element which drove the evolution of nucleic acids may also have been the structural "phenotype" of the primitive single stranded gene before becoming the translational product of the cellular double stranded gene (Mills *et al.*, 1973).

B. REPAIR OF THE PRIMITIVE NUCLEIC ACIDS

To simplify the discussion we will only consider u.v. light and assume that the sources of other possible synthesising or damaging agents were "localised" (electric discharges, radioactive minerals etc.) (see Table I). Molecules formed in their presence could have escaped their action, simply as a result of their removal away from the source.

Single-stranded oligonucleotides and polynucleotides are very susceptible to u.v. radiation damage. The removal of pyrimidine dimers by strand breakage would interrupt the continuity of the chain and lead to the destruction of template continuity and loss of information. But pyrimidine dimers can be eliminated without interrupting the chain by their photo-induced monomerisation caused by exposure to more energetic u.v. photons ($\sim$240 nm) (Setlow and Setlow, 1962), which supposedly were available in the u.v. spectrum irradiating the primitive earth (Sagan, 1965). Therefore the pyrimidine dimer type of lesions might not have exceeded a certain low proportion of adjacent pyrimidines. This physical splitting of dimers called short wavelength reversal (Setlow and Setlow, 1962) (see Fig. 2, p. 167) could only have operated until the appearance of the screening layer of ozone which cut out all short wavelength u.v. The selection of thymine and not uracil for DNA may have depended on the greater efficiency of eliminating damages to thymine by photoreversion; in uracil, a more stable hydration photoproduct formed (Lesk, 1969).

One might expect that photosensitising molecules like acetone, acetophenone, or benzophenone, etc., capable of dimerising neighbouring pyrimidines of DNA with visible light ($>$300 nm) might also be able to monomerise these dimers. Recently such monomerisation with intercalated natural indole derivatives like tryptophan or 5 hydroxytryptamine (serotonine) acting as electron donors was shown to occur with light of wavelength above 300 nm (Helene and Charlier, 1971). Partial monomerisation of some pyrimidine dimers (essentially C–C) was also obtained in *Haemophilus* transforming DNA in the presence of proflavine and short u.v. (245 and 280 nm) (Setlow and Setlow, 1965).

Bound molecules may also act as radioprotectors and have contributed to the maintenance of primitive nucleic acids: energy from u.v. or ionising radiation primarily absorbed by purines or pyrimidines migrates along the stacked bases of polynucleotide chains until trapped by intercalated (i.e. proflavine) or otherwise bound molecules (Sutherland and Sutherland, 1969; Gregoli *et al.*, 1970).

C. LIGHT-INDUCED DNA REPAIR PROCESSES

1. *Photosensitised Monomerisation of Pyrimidine Dimers, a possible "Fossil" Repair Mechanism?*

Jagger has found in microorganisms a photosensitized *non*-enzymatic repair mechanism at wavelengths *not* directly absorbed by DNA (Jagger and Stafford, 1965; Jagger *et al.*, 1970). Evidence was obtained in *Staphylococci* and *Streptomyces*, but not in *E. coli*, of a photo-induced (between 310 and 360 nm), temperature independent, dimer removal mechanism leading to a higher survival after u.v. irradiation. Similar results are also suggested from the interpretation of action spectra of *Euglena* and *Neurospora* (Ikenaya *et al.*, 1970). The understanding of such photo-induced processes both on prebiological models and in living organisms may give important leads for solving the problem of nucleic acid interrelations with other molecules during the genesis of organic material.

2. *Photoreactivation*

Primitive single stranded polynucleotides (which on account of the strong reducing conditions were probably poly-deoxyribonucleotides) could have been self-replicating and serve as coding units, such as is the case of single stranded viruses. Efficient repair mechanisms independent of the damaging u.v. light were necessary to protect the template specificity. Photoreactivation (P.R.) is an enzymatic repair system specific for pyrimidine dimers which does not require a complex chemical energy source, and which is capable of eliminating efficiently *in vivo* and *in vitro* dimers formed in single and double stranded DNA as well as in synthetic polymers (for review, see Setlow, 1966).

The photoreactivation enzyme is coded by a single gene (phr^+) in *E. coli* and in *Saccharomyces* and when purified from *E. coli* it binds to u.v. irradiated DNA independently of light. The complex is extremely temperature stable and its formation is temperature independent (Harm *et al.*, 1968); the enzyme is activated by light (>300 nm), the dimers are monomerised and the enzyme is released from the DNA (Wulf and Rupert, 1962). The chromophore of the P.R. enzyme has not yet been identified although the yeast enzyme has efficiently been purified (Minato and Werbin, 1971). The number of P.R. enzyme molecules was estimated in *E. coli* B_{S-1} by Harm (1969) and found to be of the order of 20, which may be sufficient to control dimers induced by normal sunlight.

3. *Phylogenetic Distribution*

The phylogenetic distribution of photoreactivating activity supports the idea of an early appearance of this DNA repair system. It has been found in all major groups of lower organisms, but in most cases a thorough analysis of the mechanism is still necessary. In the primitive blue-green alga *Plectonema boryanum* which may have appeared as long as 2·7 billion years ago a P.R. enzyme was extracted and assayed on *Hemophilus* transforming DNA; results are consistent with the existence of two photo-induced repair systems (probably photo-enzymatic), one of which only, using light above 400 nm, is capable of repairing the DNA of the u.v. irradiated algal phage LPP-1 (Werbin and Rupert, 1967). A photoreactivating extract has also been obtained in the alga *Anacystus* (Minato and Werbin, 1971) and in yeast, where the enzyme has been well studied (Wulf and Rupert, 1962; Cook and Worthy, 1972; Minato and Werbin, 1971).

Photoreactivating activity can be extracted from higher plants (Saito and Werbin, 1969). In *Nicotiana*, photoreactivation might be a light regulated mechanism, because when grown in absence of light, the P.R. activity is lost but can be recovered in several hours after a light induction period of only about 2 to 3 hours (Murphy and Gordon, 1971).

Photoreactivation has also been found in a wide variety of protozoa and of metazoan tissues (Cook and McGrath, 1967; Cook, 1970; Griggs and Bender, 1972). In mammals, with the exception of the evolutionary primitive marsupials (Cook and Regan, 1969) no photoreactivation is found (Trosko and Isoun, 1970).

Since a large number of phyla have not yet been examined it is premature to speculate why photoreactivation became unnecessary but this may be related to the shielding of gametes from sunlight in the higher species.

Photoreactivation is found exclusively in organelles containing DNA: DNA-free mini-cells of an *E. coli* mutant contain little or no detectable activity (Patterson and Roozen, 1972). In the case of eggs of sea urchins and amphibia, the cytoplasm shows considerable activity (Blum *et al.*, 1950; Cook and McGrath, 1967); this may be related to the high cytoplasmic DNA content and to the unability of sperm cells to photoreactivate their own DNA: sperm DNA can be reactivated by sunlight in the egg cytoplasm before forming the zygotic nucleus (Blum *et al.*, 1950). A word of caution should be added when photoreactivation is not studied enzymatically: visible light sometimes affects normal metabolism which could have secondary effects on dark repair as in *E. coli* (Jagger *et al.*, 1969) or *Schizosaccharomyces* (Fabre, 1972b).

4. *Photoreactivation of RNA*

Ultraviolet irradiated free viral RNA but not the virus (because lethal RNA-protein cross-links are probably formed) can be photoreactivated: f2 viral RNA in *E. coli* K12 (Werbin *et al.*, 1968) and TMV RNA in *Nicotiana* (Hidalgo-Salviaterra and McLaren, 1969). Plant RNA itself might also undergo P.R. (Murphy and Gordon, 1971) but the nature of the lesion and enzymatic mechanism remains to be firmly established; it is not excluded that pyrimidine hydrates are involved (Evans and McLaren, 1969) and there is no clear relationship with DNA photoreactivation.

D. JOINING OF SINGLE-STRAND BREAKS: AN ADVANTAGE OF THE DOUBLE-STRANDED STRUCTURE

A permanent double-stranded structure of DNA is not absolutely necessary for replication nor for transcription, but the complementarity of the two strands offers the possibility for a reciprocal stereochemical double check as well as for repairing single-strand breaks and for making "correc-

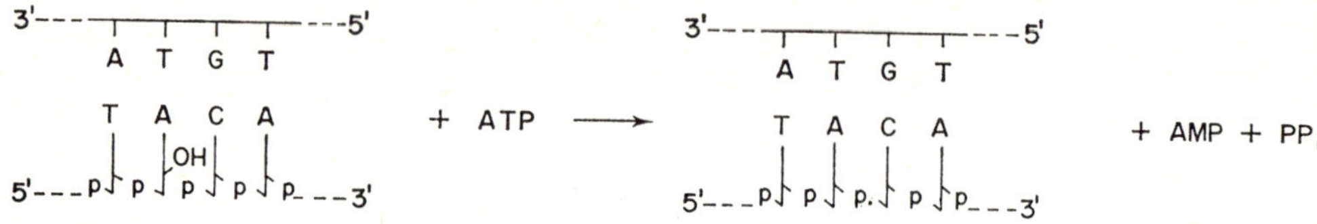

FIG. 3. Scheme for the T4 ligase reaction (from Weiss and Richardson, 1967). ATP or NAD requiring ligases probably act through an intermediate enzyme–AMP complex.

tions" involving the breakage and partial degradation of one strand in a given segment. Thus, the redundancy of a permanent double-stranded DNA structure must have had an enormous selective advantage towards the preservation of genetic information, since single-strand breakage is one of most common DNA lesions produced *in vivo* and *in vitro* by a variety of agents (section II B).

Ligases like the NAD-dependent one found in uninfected *E. coli* or the ATP dependent one induced by phage T4 (Weiss and Richardson, 1967) can seal single-strand "nicks" bearing 3′-hydroxyl and 5′-phosphate termini and re-establish strand continuity (see Gellert, 1971: Jacobs *et al.*, 1972) (Fig. 3). Their biological importance is evident from the fact that a functional DNA ligase seems to be required for the growth of phage T4 (Gellert and Bullock, 1970) and probably for recombination as well (Anraku and Lehrman, 1969; Masamune and Richardson, 1968); in any event, the ATP-dependent phage ligase can probably be replaced to a certain extent by the NAD-dependent *E. coli* enzyme (Gellert and Bullock, 1970). ATP-dependent ligases have now been isolated from a variety of organisms, including mouse embryo cells where a ligase can be induced by polyoma virus (Beard and Pitts, 1968), rabbit (Lindahl and Edelman, 1968) and human cells (Spadari *et al.*, 1971) as well as higher plants (Tsukuda and Niski, 1971).

In *E. coli*, the ligase is also necessary for DNA replication (Sugino and Okazaki, 1972; Gellert and Bullock, 1970) and repair of u.v. and X-ray damage (Pauling and Hamm, 1968; Dean and Pauling, 1970). Ligase may be the enzyme that is capable of sealing within minutes X-ray induced single-strand breaks of the type described by Sawada and Okada (1970). Single-strand breaks which would not leave the 3′OH-5′ P type of substrate for direct sealing and which may be formed by radiation or by enzymatic action (on normal or damaged DNA) would need some remodelling to yield the proper substrate (Spiegler and Norman, 1969; Brent and Wheatley, 1971), but finally, ligase has been postulated as the ultimate repair enzyme. This is also true for DNA replication and recombination (Richardson, 1969; Boon and Zinder, 1969; Clark, 1971).

The possibility of repairing double-strand breaks by ligase only would depend on the degree of stagger of the nicks in the DNA, and possibly also on the manner the broken ends could be held together by membranes or other structures (Burrel *et al.*, 1971).

E. EXCISION REPAIR

1. *Mechanism*

Only a fraction of possible DNA lesions (essentially pyrimidine dimers and some single-strand breaks) can be efficiently repaired by photoreactivation or simple ligase rejoining. Complex damage described in Section IIB (such as the destruction of the heterocyclic bases and of the sugar moiety or intra- and interstrand cross-links and cross-links with proteins and other activated molecules, etc.) could in principle be repaired by physical elimination of the damaged segment. Indeed, such a DNA repair process has been discovered in a variety of organisms. Excision repair or excision-resynthesis repair is a complex multi-step process involving the recognition of the damage, its elimination, and the reconstitution by repair replication of the strand using the intact complementary segment as template. A tentative model is shown in Fig. 4 and is described in the legend. More details can be found in other reviews (Howard-Flanders, 1968; Witkin, 1969). The mechanism of excision repair has best been studied by comparing the survival and the fate of pyrimidine dimers in wild type and u.v. radiation (*u.v.r.*) mutants of *E. coli*. The increased sensitivity of these mutants when exposed to some chemical mutagens (see Howard-Flanders, 1968) indicates that the process may also

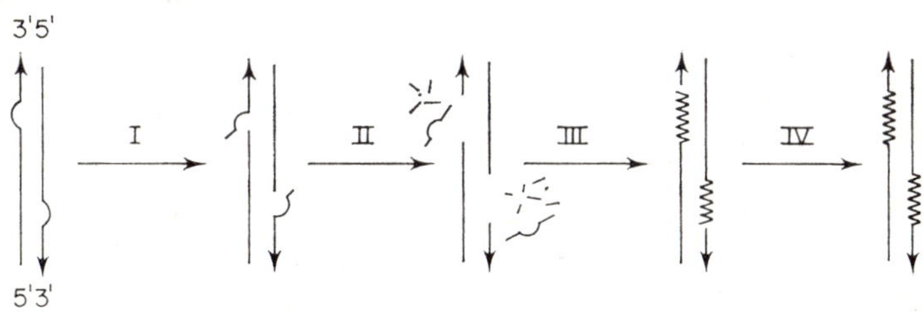

FIG. 4. Model for excision repair. ⌒ = pyrimidine dimer; — = mono- or oligonucleotides. Step I: recognition and incision by a specific "u.v." endonuclease. Step II: formation of a gap by the elimination of a pyrimidine dimer as an oligonucleotide, and by the further release of mono or oligonucleotides. Step III: repair replication. This step is probably accomplished in *E. coli* by DNA polymerase I of Kornberg. This enzyme could very well perform steps II and III simultaneously (see footnote, p. 176). In mutants deficient in this polymerase, excision gaps are filled at a lower rate, by a mechanism probably induced by u.v. and competing with the repair of other gaps formed opposite pyrimidine dimers during replication in the daughter strands (Witkin, in press). Step IV: sealing of the reconstructed strand by ligase.

repair DNA lesions other than dimers. As the *u.v.r.* mutations do not considerably increase the sensitivity to ionising radiations, the repair process controlled by this function (probably the incision step) does not seem critical.

2. *Enzymes for Excision Repair*

A detailed understanding of this sophisticated repair mechanism will soon be available since several enzymes, presumably involved in excision repair, have been isolated by Grossman and his collaborators from bacteria, such as an endonuclease specific for u.v. irradiated DNA, an exonuclease which releases pyrimidine dimers from u.v. irradiated DNA pre-exposed to the u.v. specific endonuclease (see Kaplan *et al.*, 1969; Kushner *et al.*, 1969). T4 phage directs the synthesis of an endonuclease, also specific for dimer containing DNA (Ohshikama and Sekigushi, 1970; Friedberg and King, 1971) and of an enzyme capable of removing the dimer in the nicked strand (Ohshikama and Sekigushi, 1972). The activity of such enzymes appears to be restricted to the immediate neighbourhood of the bacterial chromosome since no incision of u.v. irradiated episomal DNA could be found in *E. coli* minicells (Patterson and Roozen, 1972).

In vitro studies with DNA polymerase I have demonstrated its multiple enzymatic specificities† (Kelly *et al.*, 1969, 1970) and the biological properties of *E. coli polA* mutants clearly indicate the involvement of this enzyme in excision repair (De Lucia and Cairns, 1969; Gross and Gross, 1969). It has been suggested (Gross *et al.*, 1971; Cooper and Hanawalt, 1972) that in the absence of functional DNA polymerase I, another repair-replication system controlled by the *recA* gene of *E. coli* can partially replace it (Gefter *et al.*, 1971).

Polymerase I appears to be essential for the DNA replication of the colicinogenic plasmid E_1 (Goebel, 1972), a finding which strikingly suggests a very intimate relationship between the repair and replicating functions of the single polypeptide chain of this enzyme.

An exonuclease which has no polymerase activity, has been purified by Lindahl (1971a, b) from rabbit lungs; this enzyme efficiently liberates pyrimidine dimers *in vitro*, as part of oligonucleotides containing 5 to 8 residues. An endonuclease activity specific for u.v. irradiated DNA has also recently been detected in HeLa cells by Brent (1972).

† Kornberg's DNA polymerase I (M = 109·000) consists of a single polypeptide chain which fragments into two chains by mild proteolysis: the larger (M = 76·000) one carries a template dependent polymerase activity associated to a 3′-5′ exonuclease apparently capable of removing terminal mispaired bases or a displaced sequence having a free 3′OH; the other smaller fragment (M = 36·000) has 5′-3′ exonucleolytic activity capable of removing "frayed" 5′ ends produced by pyrimidine dimers or mismatched bases (Setlow *et al.*, 1972; Setlow and Kornberg, 1972; Brutlag and Kornberg, 1972).

3. *Phylogenetic Distribution of Excision Repair Activity*

In addition to bacteria, excision repair has been identified in yeasts by the isolation of defective mutants (Unrau *et al.*, 1971; Fabre and Moustacchi, 1972). It has also been found in human cells (Regan *et al.*, 1968; Painter, 1970; Edenberg and Hanawalt, 1972), but skin fibroblasts from patients affected with *Xeroderma pigmentosum* cannot perform the initial step of dimer excision (Cleaver, 1970), in spite of the fact that their cells apparently contain an endonuclease specific for u.v. irradiated DNA (Bacchetti *et al.*, 1972); however, in the fibroblasts of some patients this step is apparently normal (Jung, 1970); these patients are extremely susceptible to sunlight and they usually develop skin cancer. In contrast to normal human cells, rodents usually have very low excision repair activity (Trosko *et al.*, 1965; Rasmussen and Painter, 1966).

F. GENETIC RECOMBINATION AND POST-REPLICATION REPAIR

1. *Genetic Recombination as a Repair Process*

Ultraviolet light, X-rays, as well as many chemicals increase the frequency of genetic recombination in phage and bacteria (Jacob and Wollman, 1955; Roman and Jacob, 1958; Sharp and Freifelder, 1970), in algae (Lawrence, 1970), in yeasts and fungi (Manney and Mortimer, 1964; Wood and Kafer, 1969; Shanfield and Kafer, 1971; Jansen, 1970; Holliday, 1971), in *Drosophila* (Proust, 1967; Prudhommeau and Proust, 1969; Garcia Bellido, 1972) and perhaps in mouse and human germ cells (Kofman-Alfaro and Chandley, 1971). In cultured mammalian cells, u.v. and X-rays also increase the frequency of sister chromatid exchanges but the analogy between these exchanges and genetic recombination is still uncertain (Marin and Prescott, 1964; Rommelaere *et al.*, 1973).

Ultraviolet induced lesions in the DNA seem to be the cause for the increased recombination frequency. Sequences of DNA rich in adenine–thymine base pairs are expected to yield more pyrimidine dimers than regions rich in guanine and cytosine (Setlow and Carrier, 1966). Thus the finding that u.v. induced genetic exchanges in phage λ are more frequent in an A–T rich segment than in the neighbouring G–C rich segment (Cordone and Radman, 1970) suggests that pyrimidine dimers themselves are recombination-promoting sites.

2. *Ultraviolet-induced Recombination and DNA Repair*

Several phenomena from microbial radiation genetics suggest that recombination can definitely contribute to repair:

a. *Multiplicity reactivation* is a recombination dependent recovery of u.v.

irradiated phage, which is correlated with the number of damaged phage infecting the host cell (Luria, 1947).

b. *Prophage reactivation* is a recombination dependent increase in u.v. survival of phage λ when it is plated on bacteria bearing a λ prophage (Jacob and Wollman, 1955; Blanco and Devoret, 1973).

c. *Cross reactivation* or marker rescue is a rescue of u.v. damaged genetic markers from an irradiated phage by a non-irradiated phage in a mixed infection (Luria, 1952).

d. *Recombination-deficient mutants* of phage and bacteria are sensitive to u.v. and ionising radiations (Clark and Margulies, 1965; Echols and Gingery, 1968; Radman *et al.*, 1970).

The reconstruction of a "good" DNA from many damaged copies (the more, the better!) seems the simplest interpretation of the first three phenomena (Luria and Latarjet, 1947).

In fungi and yeasts a similar situation seems to occur: (a) in Ustilago heterozygotes carrying mutations in the structural gene for nitrate reductase, recombination can be detected chemically; u.v. irradiation increases the recombination frequency, but only in the surviving fraction of the irradiated cells, which suggests also that repair may be dependent on genetic recombination (Holliday, 1971); (b) some radiosensitive mutants of fungi or yeasts are recombination deficient (Jansen, 1970; Fortuin, 1971); (c) *Schizosaccharomyces Pombe* have a repair pathway which is favoured in cells with duplicated genetic material (Fabre, 1970, 1972); this effect of ploidy on radiosensitivity suggests a repair mechanism which could operate during somatic recombination and will be further discussed in Section V; (d) recombination deficient mutants of *Drosophila* have a higher frequency of induced recessive sex linked lethals (Watson, 1972).

3. *Post-replication Repair of Damaged DNA*

The radiation sensitivity of recombination deficient organisms cannot be understood as resulting from the deficiency of a simple recombination process since there is often only one copy of damaged DNA and there is nothing to be recombined. Rupp and Howard-Flanders (1968) were the first to suggest that in this case recombination-dependent repair is post-replicative and that the reconstruction of a lesion-free DNA molecule occurs by exchanges between newly replicated damaged daughter (and even cousin) DNA copies (Rupp *et al.*, 1971). Although the post-replication repair in *E. coli* and phage appears to be correlated with recombination proficiency (Rupp and Howard-Flanders, 1968; Rupp *et al.*, 1971; Radman *et al.*, 1970), there is no conclusive evidence for its identity with recombination. Rupp and Howard-Flanders and their collaborators have convincingly established that (1) newly replicated strands of u.v. damaged DNA of *E. coli* contain gaps opposite each pyrimidine dimer of the parental strand, (2) these gaps are repaired during the post-replicative DNA metabolism and post-replicative DNA repair has been mostly identified with gap-filling.

In the beautiful set of experiments involving density labelling Rupp *et al.* (1971) demonstrated u.v. enhanced molecular exchanges between the newly replicated daughter DNA molecules, but they did not show that these exchanges are prerequisite for the disappearance of gaps. In u.v. irradiated mammalian cells also, a post-replicative gap filling mechanism has been identified (Cleaver and Thomas, 1969; Klimek and Zemanova, 1971; Fujiwara and Kondo, 1972) and the post-replication sealing of gaps formed in the daughter strands does not seem to require homologous parental DNA which would be introduced by recombination (Lehman, 1972; Buhl *et al.*, 1972).

Mutations in the *recA* gene of *E. coli* which most severely abolish recombination proficiency (Clark and Margulies, 1965; Howard-Flanders and Theriot, 1966), also abolish post-replication gap-filling (Smith and Meun, 1970). On the other hand, *recB*$^-$ mutants whose recombination proficiency is reduced to less than 1% show perfect "gap-filling" (Smith and Meun, 1970) suggesting that gap-filling might be one of the numerous pleiotropic properties of the *recA*$^+$ gene product and not necessarily the consequence of recombinational events themselves.

4. *Enzymes Involved in Genetic Recombination*

RecB and *recC* genes in *E. coli* identified by Buttin code for a complex deoxyribonuclease which has recently been purified (Goldmark and Linn, 1970; Wright *et al.*, 1971). Mutation in either of these genes, in addition to reducing recombination proficiency to about 1% of the wild type level eliminates spontaneous and radiation induced DNA breakdown (Willetts and Clark, 1969). It is known that the post-irradiation DNA breakdown occurs preferentially in replicated (or replicating) segments of DNA (Hori and Suzuki, 1970); the post-replicative gaps opposite pyrimidine dimers (Rupp and Howard-Flanders, 1968) are probably the preferred substratse for the *recBC*-nuclease. All of its three *in vitro* activities† should become

† Three ATP-dependent activities of the *recBC* enzyme have been found: exonucleolytic (for single- or double-stranded DNA) and endonucleolytic (for single-stranded DNA) (Goldmark and Linn, 1970; Wright *et al.*, 1971; reviewed by Clark, 1971). The nuclease active on double-stranded DNA requires *in vitro* free DNA ends, i.e. it will not attack circular DNA (Wright *et al.*, 1971). This variety of specificities of the *recBC*-controlled nuclease *in vitro* could conceivably be reduced

apparent on DNA containing single-strand gaps. The *recA* product could directly or indirectly limit post-replicative DNA breakdown by the *recBC*-nuclease following u.v. irradiation by two processes: either by controlling the enzyme (e.g. by direct binding or by controlling the binding of something else to the enzyme) or by protecting or modifying the DNA substrate (e.g. filling the gaps); in fact, that the *recA* product controls gap-filling has already been suggested (Smith and Meun, 1970; Gross *et al.*, 1971; Cooper and Hanawalt, 1972).

Enzymes susceptible to play a role in recombination are found from procaryotes (Cassuto *et al.*, 1971) to higher eucaryotes (Richardson, 1969).

G. ULTRAVIOLET REACTIVATION

Weigle (1953) discovered that the survival of u.v. irradiated bacteriophage λ increases considerably, if plated on irradiated bacteria; simultaneously, there is an increase in the number of phage mutants induced. Jacob and Wollmann (1955) showed that there also followed an increased number of phage recombinants. This mechanism is dependent on the *recA* gene product (Miura and Tomizawa, 1968) and also on the *lex* (=*exr* gene of *E. coli* B/2) (Witkin, 1967; Defais *et al.*, 1971). Although they map apart on the *E. coli* chromosime, both *recA* and *lex* mutants exhibit similar characters:

a. radiation sensitivity and loss of u.v. reactivation as just mentioned (Miura and Tomizawa, 1968; Defais *et al.*, 1971),
b. DNA degradation following irradiation (Howard-Flanders and Boyce, 1966),
c. suppression of induced "filament" formation in *lon* strains (Donch *et al.*, 1968; Green *et al.*, 1969),
d. suppression of prophage induction (Brooks and Clark, 1967),
e. non-mutability by u.v. and X-rays (Witkin, 1967).

Survival of irradiated λ on *lex*$^+$ or *lex*$^-$ non-irradiated hosts is identical (Defais, Fauquet and Radman, unpublished) while if one irradiates host cells with u.v. doses optimal for mutagenesis and for prophage induction, there is a strong increase in survival on the *lex*$^+$ but not on *lex*$^-$ bacteria, an effect which is strongly masked if the bacteria are photoreactivated or treated immediately by an inhibitor of protein synthesis (Ono and Shimazu, 1966). This indicates that the pyrimidine dimers formed during irradiation possibly

---

in vivo, if ATP were required for local opening of the double helix in order to make the single stranded substrate suitable for nicking and possibly subsequent exonucleolytic digestion. The fact that the activity of this nuclease causes an increased spontaneous and u.v. induced DNA breakdown in *recA*$^-$ mutants suggests that the *recA* product could be either a protein protecting DNA sites sensitive to *recBC* nuclease or a specificity-determining element for this nuclease which itself has a rather devastating effect on DNA *in vivo* and *in vitro*.

induces the *lex* gene product. Ultraviolet reactivation could then be explained if this product increased the *recA* control on the *recBC* nuclease, rendering post replicative gaps less vulnerable to this enzyme; these gaps would then be filled by an error prone (mutagenic) process.

The induction by u.v. light or chemicals of a function is not an unusual proposition: these agents can increase the synthesis of a DNA polymerase in *Tetrahymena* (Westergaard and Pearlman, 1969; Keiding and Westergaard, 1971). In *Tetrahymena* also, Calkins (1967) has shown the activation of a repair mechanism upon substantial u.v. or X radiation injury, only when the biochemical apparatus for protein synthesis is functional. Holliday (1971) has suggested that one or more of the enzymes involved in the recombination repair mechanism may be u.v. induced in *Ustilago*. In *Saccharomyces* there is a reversal of X-ray injury by subsequent exposure to u.v. (Calkins and Todd, 1968). In mammals also, a u.v. irradiated virus has a better chance of surviving when plated on u.v. or X-ray treated cells (Bockstahler and Lytle, 1971). This suggests that a mechanism similar to u.v. reactivation which probably operates by the induction by radiation of a repair pathway, is still operative in higher organisms.

H. SUBSTRATE SIZE FOR VARIOUS DNA REPAIR SYSTEMS; EVOLUTIONARY IMPLICATIONS

In the preceding discussion we have postulated an evolutionary sequence for the appearance of DNA repair mechanisms; short wavelength reversal was followed by photoreactivation after which came excision repair, recombination and perhaps another less well understood mechanism, post-replication repair. In summary, there is good evidence suggesting that in the very early stages of prebiotic evolution, u.v. light of wavelength around 240 nm played an important role; its subsequent screening by ozone (Sagan, 1965; Ponnamperuma *et al.*, 1963) rendered this mechanism inoperative. Photoreactivation by near u.v. is found in all groups of lower organisms but not in higher mammals (Cook and McGrath, 1967). Excision and post-replication repair are in contrast probably universally distributed from bacteria to man and they can obviously handle a much wider variety of damage.

Besides, the phylogenetic distribution and the degrees of complexity of enzymatic systems and of energy requirements, the nature and size of the minimal substrate needed for each DNA repair system suggests a possible evolutionary pathway. Short wavelength reversal can repair a single pyrimidine dimer induced in a dinucleotide (Setlow and Setlow, 1962). Photoreactivation probably requires the minimum length of nine nucleotides (Setlow and Bollum, 1968), no matter whether they belong to a double- or single-stranded structure, to natural or synthetic polymer. Excision repair requires high molecular weight, double stranded DNA. Considering the mechanism of excision repair (see Fig. 4) and the fact that about ten to fifty neighbouring nucleotides can be excised by *E. coli in vivo* with one single dimer (Billen *et al.*, 1967; Ley and Setlow, 1972), only a large duplex macromolecule could undergo such repair. The post-replication gaps in u.v. irradiated DNA may be about thousand nucleotides long (Rupp and Howard-Flanders, 1968; Lehmann, 1972; Buhl *et al.*, 1972) and the hybrid overlap region formed during genetic recombination is also of about one thousand nucleotide pairs long (see Meselson, 1967); thus genetic recombination is likely to require double-stranded DNA even larger than the minimum substrate for excision repair. Yet one must keep in mind that there may exist several different modes of genetic recombination (see Clark, 1971).

I. COORDINATION OF DNA REPAIR AND REPLICATION

Why do excision repair and post-replication repair co-exist in the cells from bacteria to man? Why is the presence of both systems necessary for efficient DNA repair? If both repair systems are independent why was not the more efficient one exclusively selected? Whatever the detailed mechanism and structure of the replication fork, we must assume the necessity for the physical continuity of the two parental DNA strands in the sequence being replicated. Breaks and gaps in the parental strands would lead to an abortive replication which should eventually be lethal. Therefore the structural requirements for the replication of u.v. damaged DNA must be either to complete excision repair by establishing the final covalent bond immediately ahead of the replication fork or not to start to excise pyrimidine dimers proximal to the replication fork. In the latter case a gap will be formed opposite to dimer in the daughter strand and filled up afterwards (Rupp and Howard-Flanders, 1968). If so, the efficiency of excision repair alone should increase by delaying DNA replication while it is expected to decrease the efficiency of post-replication repair (or recombination-dependent repair), which requires DNA replication. These predictions have been proved correct by genetic experiments (Radman *et al.*, 1970; Radman and Errera, 1970). Any treatment liable to delay DNA synthesis, like "liquid holding" is expected to influence this coordination between repair mechanisms. In *E. coli* one or two pyrimidine dimers are lethal to cells which have lost both repair systems, twenty or fifty dimers are lethal when excision or *recA* dependent repair operates alone, but it takes more than 3000 dimers to kill cells having normal repair (Howard-Flanders *et al.*, 1968); this shows the great efficiency of the coordination between pre- and post-replication repair mechanisms since the efficiency of the two systems working together is more than additive. Theoretical interpretation of the experimental data (Fornili *et al.*, 1971) suggests that this cooperativity arises from their coordination with DNA replication. As incision of the DNA chain close to the base damage by ionising radiation does not seem critical, this type of coordination may be limited to pyrimidine dimer type of lesions.

IV. Elements of Variability of Genetic Information

A. SPONTANEOUS MUTATIONS

1. *Preliminary Remarks*

Ultraviolet light was probably the most ubiquitous environmental factor, before it became screened by ozone, for inducing variation in polynucleotide chains; but the molecular mechanisms of either spontaneous or radiation induced mutagenesis are still completely obscure. One may only make a certain number of hypotheses on the basis of what is known of the response of DNA to various external agents, physical or chemical, or from organisms whose spontaneous mutation frequency has been well studied. Growing evidence seems to indicate that important mutagenic events occur during replication of DNA (Haas and Doudney, 1960; Lieb, 1960; Cox, 1970; Drake, 1970) and that some may depend on recombination or repair. We will attempt in this section to emphasise these relationships and indicate briefly the type of changes in the primary DNA structure which is expected to occur as a result of the action of a variety of physical or chemical agents. An excellent discussion has been given by Drake (1970).

2. *Functions Involved in Spontaneous Mutagenesis*

In *E. coli* most genes (*recA*, *lex*) known to influence some radiation induced mutation processes (*vide infra*) do not seem to greatly affect spontaneous mutation rates (Witkin, 1969) although there may be a slight dependency on *recA* for the reversion of arginine (*ArgF*) amber mutants (Kondo *et al.*, 1970). There is no clear indication either that the *polA* gene (coding for DNA polymerase I active in repair replication) is necessary for spontaneous mutations (Witkin, 1971; Smirnov *et al.*, 1971; Caillet-Fauquet *et al.*, 1972) although there again may be some exceptions (Kondo *et al.*, 1970) depending on the nature of mutations in this gene.

In phage λ, spontaneous clear plaque mutations significantly depend on the phage recombination function *red*† (Defais *et al.*, 1971). Mutation rates may also depend on whether the phage is replicating as part of the bacterial chromosome or not: in the lytic condition the rate of reversion of amber mutants may be increased as much as a hundred-fold (Dove, 1968).

In phage T4, Drake (1970) has found that spontaneous mutations in the *rII* locus may be independent of DNA replication: if stored at 20°C, the overall spontaneous mutation rate is of 0·1 mutation per phage per year, and ten times less at 0°; changes are chiefly base transitions,‡ a few frame-

† *Red* is a complex phage coded function responsible for autonomous vegetative recombination of phage λ (Signer and Weil, 1968; Echols and Gingery, 1968).

‡ Replacement of A by G or C by T or vice versa.

shifts, possibly some transversions† and no deletion. But during replication, frameshifts are by far the greatest mutational event (0·7) with a few transitions (0·2) and few transversions and deletions (0·1). What is most striking is the tremendous spontaneous mutability of some loci (hot spots) (Benzer, 1961): the base sequence in one of these loci has been analysed by Okada *et al.* (1972) who have found that it contains six adjacent A–T base pairs.

In yeasts, Magni (1964, 1969) was the first to demonstrate that spontaneous frameshift mutations occur during meiotic interallelic recombination (Magni, 1969) and similar results were obtained in *Schizosaccharomyces Pombe* (Friis *et al.*, 1971) but there does not appear to be a dramatic dependency of spontaneous mutagenesis on radiation sensitivity functions (Nasim, 1968).

Summarising, in bacteria and phage, there is increasing evidence that the DNA replicating machinery is a decisive element in determining spontaneous mutation rates (Drake, 1970; Cox, 1970), but the contribution of repair functions is much less clear. In yeasts, on the other hand, there is a good correlation between spontaneous frameshift mutations and meiotic recombination.

B. MUTATOR GENES

The mutator gene (*mut* T) of *E. coli* (Treffers *et al.*, 1954) increases the spontaneous mutation rate by about one thousand-fold and it produces AT→GC transitions in the A chain of tryptophan synthetase at a rate such that a small increase in the buoyant density of the total DNA can be measured after 1000 generations; this would indicate an increase in its GC content by as much as 0·3% (Cox and Yanofsky, 1967).

However, mutator genes have not yet been exploited to study how this type of mutation occurs. The mutator gene of *E. coli* affects phage λ only during its replication (Cox, 1970), but it has little effect on the mutation rate of phage T4. This is related to the fact that this phage codes for its own replicative DNA polymerase (gene 43); some mutations in this gene produce a considerable increase (up to a thousand-fold) of the spontaneous mutation rate. This suggests that the polymerase plays a direct role in the selection of the correct base during DNA replication (Drake *et al.*, 1969; Drake, 1970). Mutator genes, although less well analysed, have also been found in yeasts (von Borstel *et al.*, 1968) and in human leukemic cells (Springate and Loeb, 1973).

C. RADIATION INDUCED MUTAGENESIS

1. *Functions Involved in Radiation-induced Mutagenesis*

In *E. coli* two classes of mutants are known to abolish u.v. radiation induced mutagenesis: *lex* (Witkin, 1967) and *recA* (Witkin, 1969b; Miura and Tomi-

† Replacement of a purine by a pyrimidine or vice versa.

zawa, 1968). Mutations in the *recB* and *recC* do not impair u.v. mutagenesis (Miura and Tomizawa, 1968). With X-rays and many chemicals, mutations cannot be obtained in *recA*⁻ strains (Kondo *et al.*, 1970) nor in *lex*⁻ strains (Bridges *et al.*, 1968). In yeasts there are some mutants sensitive both to u.v. and X-rays which behave like *recA* or *lex* bacterial mutants (Averbeck *et al.*, 1970). The *lex* function is probably induced by u.v.: this suggests that the mutagen itself is capable of inducing a repair mechanism which increases the frequency of the final mutagenic events.

The functional state of the organism may also be critical: when the histidine operon of *Salmonella* is derepressed, the rate of u.v. induced mutations in this operon is increased several fold; as this is not affected by the *hcr* mutation (probably affecting excision repair) transcription might in some way interfere with another repair mechanism (Savić and Kanazir, 1972).

2. *Chemical Changes During Radiation-induced Mutagenesis*

Thymine dimers can be induced specifically by photosensitised illumination in phage T4; they induced deletions (0·1), transitions chiefly of the GC→AT type (0·33) (which indicate that the mutation does not necessarily occur opposite a dimer), frameshifts (0·37) and less well characterised transversions (0·12) (Meistrich and Drake, 1972). Ultraviolet induced mutations of *E. coli* in the A chain of the tryptophan synthetase (Yanofsky *et al.*, 1966) are 80% single base substitutions (transitions or transversions), the rest being frameshifts (Witkin, 1969). In *E. coli* X-rays, as well as thymine starvation, probably induce AT→GC transitions (Bridges *et al.*, 1968) and in *Neurospora* among the 60% mutations which could be analysed, half were frameshift mutations, the other half base pair substitutions (Malling and de Serres, 1967). Furthermore, X-rays can occasionally induce gene duplications like those observed in *Drosophila* sperm (Slizynska, 1969).

The chief mutagenic signals induced by u.v. are the unexcised pyrimidine dimers: after replication they temporarily leave gaps in the daughter strands, which in *E. coli* are finally filled by a *recA* and lex dependent error prone process. X-ray induced single strand breaks are likely to be a good substrate for nucleases; these might uncover single stranded regions leaving gaps, the filling of which would be mutagenic (Bridges *et al.*, 1968) but base damages could also be mutagenic.

D. CHEMICAL MUTAGENESIS

We will not go into much detail concerning this problem, crucial however for the understanding both of spontaneous and radiation induced mutations.

1. *Base Starvation and Base Analogs*

Thymine starvation is mutagenic in thymine requiring strains of *E. coli*

(Kanazir, 1958). In these circumstances single-strand gaps are formed in bacteria, presumably in thymine rich sequences (Menningmann and Szybalski, 1962; Freifelder and Katz, 1971); it is probable that such gaps also occur in starved *E. coli* and their repair might be the *lex* dependent mutagenic step (Bridges *et al.*, 1968) leading to AT→GC transitions. Pietrszykowska (1973) has shown that BUDR mutations in phage λ are also *recA* and *lex* dependent, which suggests that they do not occur through simple mispairing.

2. *Carcinogens and Alkylating Agents*

In *E. coli*, the mutagenic effect of some carcinogens and alkylating agents (Drake, 1969) are dependent on *recA* and *lex* genes, some on excision (*u.v.r.*) genes and some independent of repair processes (Witkin, 1969; Kondo *et al.*, 1970). In the same organism, Cerda Olmedo *et al.* (1968) have shown that nitrosoguanidine induced mutations occur in the region of the replication fork. At present it is not possible to integrate any of these data with what is known for radiation induced mutagenesis until the fate of the damage to the

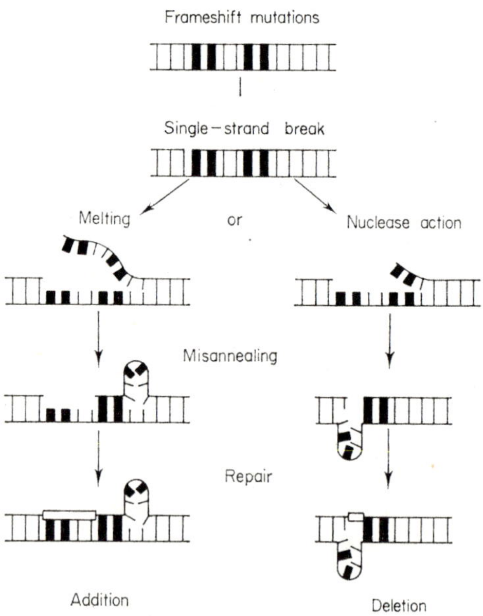

FIG. 5. Hypothetical mechanism for the production of frameshift mutations by misrepair after Streisinger *et al.* (1966). The heavy bars in the schematic DNA molecules represent short regions of base pair homology; the open line segment at the bottom represents newly synthesised DNA. The addition mutant has gained four bases, resulting in a reading frameshift to the left; the deletion mutant has lost four bases resulting in a reading frameshift to the right (figure and legend from Drake, 1970, p. 144).

bases has been analysed. The data clearly indicate that some repair pathways are shared for eliminating damage by chemical and physical agents and they also suggest that related mutation mechanisms probably exist.

3. *Intercalating Agents*

Acridine dyes efficiently cause frameshift mutations (base additions or deletions) in the *rII* gene of phage T4 (Barnett *et al.*, 1967; Drake, 1969). In the lysozyme gene of the same phage, proflavine may cause the addition of up to four or five base pairs as a result of a single event as suggested from amino acid sequence analysis (Okada *et al.*, 1972). Lerman (1963) suggested that by intercalating in repetitive sequences they might induce illegitimate crossing over, accompanied by the subsequent addition or deletion of bases; this would be consistent with the idea that frameshift mutations occur during genetic recombination. In phage T4, however, acridine mutations appear not to arise during recombination (Lindstrom and Drake, 1970). The dye might interfere with repair and stabilise unpaired regions in highly repetitive sequences (Fig. 5) as suggested by Streisinger (Streisinger *et al.*, 1966). In *E. coli*, proflavine efficiently induces mutations only during conjugation and here they occur in the vicinity of recombinational events (Sesnowitz-Horn and Adelberg, 1968). In *Saccharomyces*, like spontaneous frameshift mutations (Magni, 1969), acridine induced mutations occur during meiosis but seldom during mitosis and the mutants have a high probability of being recombinant for outside markers (Magni *et al.*, 1964).

E. GENETIC RECOMBINATION, MUTAGENESIS AND DNA REPAIR

The above considerations indicate that in *E. coli* (but no generalisation is at present possible) induced mutagenesis is often dependent on the *recA* function, but this does not mean that mutations necessarily depend on genetic recombination. The elegant experiments of Rupp *et al.* (1971) demonstrate, as we have already stated, that frequent exchanges occur between the daughter chromosomes following u.v. irradiation; they do not imply that recombination is the primary mechanism of reconstitution of gap-free DNA. Since the *recB* mutants show normal "gap-filling "repair (Smith and Meun, 1970) and only 1 % recombination proficiency, we would like to suggest that *recA* is involved in two post-replicative repair mechanisms: first in an induced mutagenic gap-filling process and second in a recombination repair mechanism involving exchanges between the daughter DNA molecules which need not cause mutations. These two *recA*-dependent DNA repair mechanisms can be demonstrated using phage λ assays. Post-replication *recA* and *red* dependent recombination repair (in the strict sense) of the u.v. irradiated phage λ occurs also in lex^- bacteria (Radman *et al.*, 1970; Defais, Fauquet, Radman, unpublished). This repair has a low capacity (i.e. it is not efficient on highly damaged DNA) and does not lead to any important increase in mutagenesis. On the other hand, the presumed gap-filling repair (which has not yet been demonstrated in λ phage) occurs only in u.v. irradiated *lex*$^+$, *recA*$^+$ *E. coli*; this repair, which is induced in the host cell (u.v. reactivation) is very efficient and highly mutagenic. Mutations could be introduced by this mechanism during repair of gaps produced in DNA either after excision of a fraction of the dimers or in daughter strands opposite unexcised dimers (Witkin and George, in press).

The relationship between functions involved in genetic recombination and mutagenesis found in prokaryotes is probably similar in eukaryotes, but here the situation is much less clear. In *Saccharomyces*, mutants, resembling *lex* mutants of *E. coli*, defective in u.v. induced mutations have also been isolated and these are moderately u.v. sensitive and X-ray sensitive; their meiotic and radiation induced mitotic recombination rates are not reduced (Lemontt and Mortimer, 1970; Lemontt, 1971). Other strains behave phenotypically like *u.v.r. E. coli* (excision defective) and u.v. induced mutation rates are increased; this has been found both in *Schizosaccharomyces Pombe* (Nasim, 1968) and in *Saccharomyces cerevisiae* (Averbeck *et al.*, 1970). Caffeine, which seems to inhibit specifically post-replication repair in eukaryotes, decreases u.v. mutagenesis in *Schizosaccharomyces Pombe* (Loprieno, 1971). In *Aspergillus*, strains with an increased mutagenic activity have been isolated: some have an increased frequency of spontaneous intragenic mitotic recombination but others are deficient for mitotic recombination (Jansen, 1972).

In mammals the situation is still less clear. Like in yeasts, caffeine also diminishes the frequency of u.v. induced mutations (Trosko and Chu, 1971). There also may be regions of the chromosomes more or less susceptible to X-ray (Lifschyts and Falk, 1968, 1969), BUDR or hydroxylamine (Engel *et al.*, 1967), dibenzanthracene or even Rous Sarcoma virus (Kato, 1967, 1968) induced aberrations, suggesting that all regions of the chromosome are not equally susceptible (hot spots?).

These data on eukaryotes are extremely indirect and they are mentioned only to draw attention to facts which do recall to some extent similar ones found in phage and bacteria, and which might indicate the existence of mechanisms likely to be "universal".

V. Increase in Genetic Information and Cellular Radiosensitivity

A. TARGET SIZE AND DNA REPAIR

The quantity of DNA in a cell can increase as the result of several events: by the appearance of redundant sequences either in "parallel" like when ploïdy is increased or in polytheny, or in "series" by duplication of existing DNA sequences, which could result from illegitimate crossing over between identical DNA molecules (in this case leading to redundancy of information in one progeny molecule and to a deletion in the other) (see Ohno, 1970). The quality of information may improve as a result of viable mutations leading to the increased efficiency of a gene product and in the case of duplicated sequences, to the divergent evolution of *t*RNAs (Holmquist *et al.*, 1973) of related proteins as is suggested by the amino acid sequences of myoglobin and the haemoglobins (see Ohno, 1970) or by the "pedigree" of protamines as postulated by Fitch (1971); it could also improve as a result of crossing over between the DNA of related individuals. A simultaneous quantitative and qualitative increase of information could occur by the introduction of a

TABLE II

Genome size, u.v. light and cell survival

	Genome size (daltons)	D_{37} ergs/mm^2	Pyrimidine dimers for D_{37}	References
Phage λ	$3 \cdot 10^7$ (1)	810 (1)	38·8 (1)	(1) Radman *et al.*, 1970
E. coli K12	3×10^9 (2)	500 (2)	$\sim 3 \times 10^3$ (2)	(2) Howard-Flanders, 1968
Saccharomyces cerevisiae				(3) Petes and Fangman, 1972
n	$\sim 10^{10}$ (3)	$\sim$ 800 (4)	$\sim 3{\cdot}5 \times 10^4$	(4) Averbeck *et al.*, 1971
		$\sim$ 800 (5)	(assuming D_{37}	(5) Resnick, 1969
		$\sim$ 1200 (6)	1000 ergs/ mm^2) (8)	(6) Cox and Parry, 1968
				(7) Parry, 1971
2n		$\sim$ 1800 (4)		(8) Unrau *et al.*, 1971
		> 2000 (5)		
		$\sim$ 2000 (7)		
Schizosaccharomyces Pombe				(9) Fabre, 1972
n		1800 (9)	$\sim 13 \times 10^4$ (10)	(10) Fabre and Moustacchi (pers. comm.)
2n		$\sim$ 1900 (9)		
Chinese hamster	3×10^{12} (11)	$\sim$ 70 (13)	$\sim 16{\cdot}5 \times 10^5$ (15)	(11) Huberman and Riggs, 1966
				(12) McCarthy, 1969
Human fibroblasts	$3{\cdot}6 \times 10^{12}$ (12)	$\sim$ 30 (14)	$\sim 7{\cdot}5 \times 10^5$ (16)	(13) Cleaver, 1969
				(14) Cleaver, 1970
				(15) Cleaver and Trosko, 1969
				(16) Cleaver and Trosko, 1970

foreign genome as in lysogeny, transduction by phages or transformation of cells of higher species by viruses (Hedge, 1972; Mourant, 1971).

Evolution depended therefore to a great extent on recombination and mutation, the rate of which can be increased by a variety of physical and chemical agents which could, like u.v., have played a role in accelerating the natural processes but which would have been of no evolutionary use without efficient mechanisms to eliminate the damage they induce in DNA.

Physicists have for a long time considered that radiations inactivate certain cellular targets, a concept which culminated in Lea's classical work (1946). If one admits that DNA is one of these targets and considering that during evolution the DNA content of cells increased, one should expect their radiosensitivity to increase at the same time. Since the discovery of repair mechanisms, the target theory has been severely criticised; however, radiosensitivity should still be proportional to target (DNA) size if repair is taken into consideration, i.e. in absence of or with constant repair. The comparison of survival to u.v. light of phage λ carrying a deletion in a region not required for vegetative growth, shows a direct correlation between sensitivity and amount of DNA (Radman, 1971). This suggests that in this case lethality occurs at the level of DNA replication (when it is blocked or abortive, DNA breakdown is known to occur).

The probability of such an event is expected to increase with the quantity of DNA. The increase in size of the replicating material is therefore an intrinsic factor of radiosensitivity, although this may easily be masked by various degrees of repair. If one compares the radiosensitivity of cells during evolution, one does not find the increase in radiosensitivity expected from the enormous increase in amounts of DNA: a mammalian cell contains 10^3 times more DNA than *E. coli*, but its radiosensitivity to u.v. light is only slightly greater (see Table II) and not more than 10 to 100 times greater in the case of X-rays (Lehman, 1972).

B. SEXUALITY: AN ADVANTAGE IN OVERCOMING DNA HAZARDS

The selective advantage of a permanent double-stranded over a single-stranded informative molecule has already been emphasised in relation to DNA repair. The diploid state resulting from sexual conjugation was obviously also, as suggested by the consideration of Terzi (1961) and of Kaplan and Moses (1964), of tremendous selective advantage, on one hand as a means of having in a single cell two sets of chromosomes carrying each the information on one DNA strand and its complement on the other and on the other hand as an increased possibility for recombination. The evolution towards a long diploid part of the life cycle is obvious and clearly examplified in plants where the duration of the diploid phase becomes progressively dominant over the haploid.

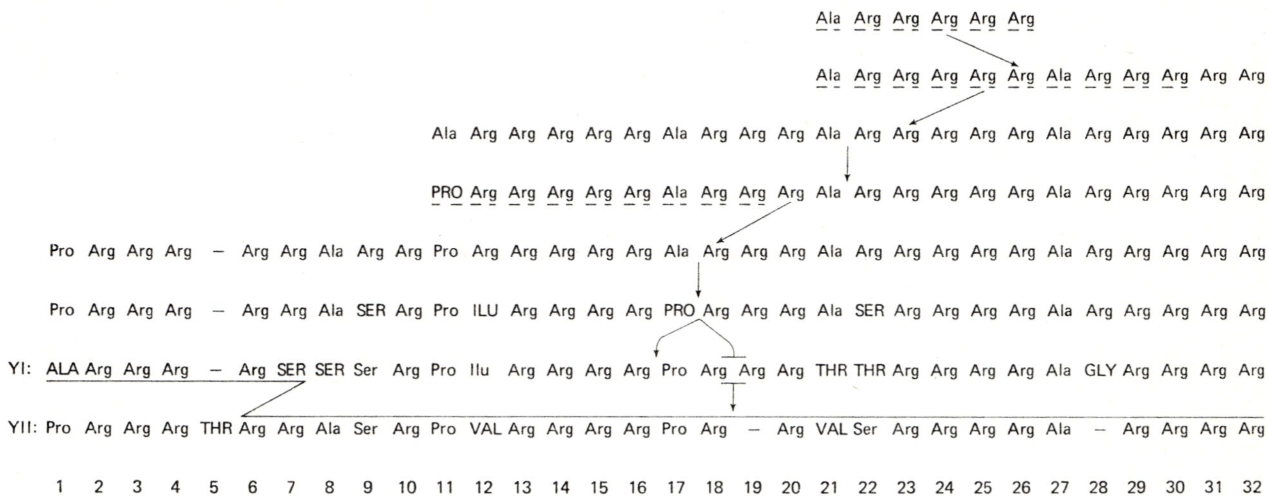

FIG. 6. An evolutionary pathway for three clupeins (Y_I, Y_{II} and Z), the protamines of a Pacific herring (Fitch, 1971). From the proposed ancestor peptide a series of gene duplications (– – –) and point mutations (capitals) could lead to Y_I and by addition of Thr 5 and by other point mutations to Y_{II}. Recombination between the genes coding Y_I and Y_{II} could lead to the gene coding for clupein Z(——). From the codes for ala (GCX) and arg (CGX or AG Pu) one can imagine an ancestor DNA with near to alternate GC sequence—which would make it extremely resistant to u.v. light; the high arginine content of the ancestor protein would favour the formation of a complex with DNA which would stabilise it.

A direct implication of a repair mechanism related to mitotic recombination is suggested from the effects of ploidy on cell survival to u.v. light: in *Schizosaccharomyces Pombe*, the higher resistance of the diploid cells compared to haploid is related to a repair mechanism more efficient in the diploid strain; this mechanism is lost in some radiation sensitive strains, it is inhibited by caffeine which also inhibits mitotic crossing over and is interpreted as a recombination dependent repair (Fabre, 1970, 1972). The radioresistance of cells does not continue to increase with their ploidy in spite of the increased possibility of complementation and recombination; in yeast this is probably due to an increased frequency of dominant lethal hits, which could result from illegitimate chromosomal rearrangements which become more probable as ploidy increases (Mortimer, 1958).

In mammalian cells, it has been suggested that exchanges between homologous chromatids or chromosomes could favour a true recombination repair mechanism, which would compensate the expected increase of dominant lethal events (Rommelaere and Errera, 1972; Rommelaere *et al.*, 1973).

One could also suggest that meiotic crossing over participates in DNA repair and plays a fundamental role in the elimination of lesions. In oocytes which remain blocked in meiotic prophase during long periods of time, one expects damages to accumulate and it is not unreasonable to think that meiotic recombination might constitute a mechanism of eliminating such damage in addition of course to the advantage that recombination itself has for increasing genetic variability. Repair synthesis and associated enzymatic activities have been found during the pachytene state in Lilium (Hotta and Stern, 1971; Howell and Stern, 1971); an increased frequency of meiotic crossing over occurs in *Drosophila* after u.v. (Prudhommeau and Proust, 1969) or X-rays (Leigh, 1969) and in X-irradiated *Chlamydomonas* (Lawrence, 1970). Unscheduled DNA synthesis increases during the chiasmata stage of meiosis in mammalian germ cells treated with u.v. or X-rays (Kofman-Alfaro and Chandley, 1971; Chandley and Kofman-Alfaro, 1971) suggesting that repair of damage may occur as well as recombination strand resynthesis.

C. GENOME SIZE, DNA REDUNDANCY, IN RELATION TO REPLICATION AND REPAIR

If one compares survival to u.v. of haploid yeasts and *E. coli*, in spite of their having four times more DNA, yeasts are more resistant. In *E. coli*, as well as in phage λ, the u.v. resistance of the wild type species was shown to be the consequence of the coordination between prereplicative excision repair (*u.v.r.* genes) and post-replicative *recA* dependent repair (Section III G). A similar coordination appears to hold also in yeasts; repair mutants responding to u.v. light like *u.v.r.* or *recA* defective mutants of *E. coli* have been isolated in *Schizosaccharomyces* (Fabre, 1970) and *Saccharomyces* (Khan *et al.*, 1970): the wild type supports considerably more damage than either of the mutants having only one repair system. If one passes from the diploid yeast to the mammal, the increase in sensitivity of the latter is not proportional to the increase in genome size (Table II). It would be tempting to postulate that such a coordination between repair systems and DNA replication has been perpetuated through evolution. If this were the case, the subdivision of the genome into individual chromosomes and even many more smaller replicons (Huberman and Riggs, 1968; Callan, 1972) may have favoured such a coordination.

Another aspect concerns the functional inactivation by lethal mutagenic lesions in coding DNA. Redundancy of DNA sequences in series or in parallel constitutes a mechanism capable of releasing the selective pressure when a redundant informational unit is damaged (Britten and Kohne, 1969). The brilliant analysis of spontaneous mutagenesis in mammals by Ohta and Kimura (1971) can be adopted here; their theoretical method was so powerful that they could estimate the degree of DNA redundancy from the discrepancy between the "real" mutation rate per nucleotide and the observed mutation rate. If, according to them, only about 1% of the mammalian genome is actively coding, the rest being formed of tandem repetitions and "spacers", then the functional mammalian DNA target is just about ten times bigger than that of *E. coli*.†

The lesions in non-coding mammalian DNA sequences, even if they were highly mutagenic, might be of no great harm to the functional cell machinery, granted the known efficient post replicative repair (Lehmann, 1972). On the other hand, they will have a drastic effect on the whole replicon since these present a direct obstacle for DNA replication; this could be compensated by a greater efficiency of repair.

In order to estimate the effect of the organisation of the genetic material (chromosomes, replicons, repetitions, etc.) on cellular resistance to DNA damage, one should for instance compare *E. coli* and haploid yeast cells both totally deficient in DNA repair. If, in spite of its four-fold higher DNA content, such mutant yeast still shows higher resistance to u.v., we may assume that "silent" DNA is likely to have contributed to that resistance. If such a repairless yeast becomes more sensitive than a repairless bacterium, then it would be certain that the efficiency of yeast DNA repair systems and/or their coordination, perhaps related to the formation of many chromosomes and replicons, is better than in bacteria.

† The estimated number of "genes" in man is about 3×10^4 (Muller, 1967; Ohta and Kimura, 1971) which is roughly ten times more than the presumed number of bacterial genes, where no gene repetitions are found.

Acknowledgments

We thank Dr. E. Moustacchi and Dr. R. Thomas for helpful discussions.

This work was carried out under contract Euratom-ULB (099-72 1BIAB) and with the help of an agreement between the Belgian Government and the Université libre de Bruxelles concerning priority action for collective basic research.

References

Akaboshi, M., Kawai, K. and Waki, A. (1971). *Biochim. biophys. Acta* **238**, 5–7.
Anraku, N. and Lehrman, I. B. (1969). *J. molec. Biol.* **46**, 467–479.
Averbeck, D., Laskowski, W., Eckardt, F. and Lehmann-Brauns, E. (1971). *Molec. Gen. Gent.* **107**, 117–127.
Bacchetti, S., Vanderplas, A. and Veldhuisen, G. (1972). *Biochem. biophys. Res. Commun.* **48**, 662–669.
Barnett, L., Brenner, S., Crick, F. H. C., Shulman, R. G. and Watts Tobin, R. J. (1967). *Phil. Trans. R. Soc. Ser. B* **252**, 487–560.
Beard, P. and Pitts, J. D. (1968). *Biochem. J.* **110**, 48 P.
Benzer, S. (1961). *Proc. natn. Acad. Sci. U.S.A.* **47**, 403–415.
Billen, D., Hewitt, R. R., Lapthistophon, T. and Achey, P. M. (1967). *J. Bacteriol.* **94**, 1538–1545.
Bishop, M. J., Lohrmann, R. and Orgel, L. E. (1972). *Nature, Lond.* **237**, 162–164.
Black, S. (1971). *Biochem. biophys. Res. Commun.* **43**, 267–272.
Blanco, M. and Devoret, R. (1973). *Mutat. Res.* **17**, 293–305.
Blum, H. F., Robinson, J. C. and Loos, G. M. (1950). *Proc. natn. Acad. Sci. U.S.A.* **36**, 623–627.
Bockstahler, L. E. and Lytle, C. D. (1971). *J. Virol.* **8**, 601–602.
Boon, T. and Zinder, N. D. (1969). *Proc. natn. Acad. Sci. U.S.A.* **64**, 573–577.
Brendel, M., Khan, N. A. and Haynes, R. H. (1970). *Molec. Gen. Gent.* **106**, 289–295.
Brent, T. P. (1972). *Nature, New Biol.* **239**, 172–173.
Brent, T. P. and Wheatley, G. A. (1971). *Int. J. Radiat. Biol.* **19**, 339–348.
Bridges, B. A., Law, J. and Munson, R. J. (1968). *Molec. Gen. Genet.* **103**, 266–273.
Britten, R. J. and Kohne, D. E. (1969). *In* "Handbook of Molecular Cytology" (ed. Lima de Faria, A.), pp. 37–51. North Holland, Amsterdam.
Brooks, K. and Clark, A. J. (1967). *J. Virol.* **1**, 283–293.
Brutlag, D. and Kornberg, A. (1972). *J. biol. Chem.* **247**, 241–248.
Bugg, C. E., Thomas, J. M., Sundaralinjam, M. and Rao, S. T. (1961). *Biopolymers* **10**, 175–219.
Buhl, S. N., Setlow, R. B. and Regan, J. D. (1972). *Int. J. Radiat. Biol.* **22**, 417–424.
Burnotte, J. and Verly, W. G. (1971). *J. biol. Chem.* **246**, 5914–5918.
Burrel, A. D., Feldschreiber, P. and Dean, C. J. (1971). *Biochim. biophys. Acta* **247**, 38–53.
Buttin, G. and Wright, M. (1968). *Cold Spring Harb. Symp. quant. Biol.* **33**, 259.
Caillet-Fauquet, P. and Defais, M. (1972). *Mutat. Res.* **15**, 353–355.
Calkins, J. (1967). *Int. J. Radiat. Biol.* **13**, 283–288.
Calkins, J. and Todd, W. (1968). *Int. J. Radiat. Biol.* **14**, 487–491.
Callan, H. G. (1972). *Proc. R. Soc. Ser. B* **181**, 19–41.

Calvin, M. (1969). *In* "Chemical Evolution". Clarendon Press, Oxford.
Cassuto, E., Lash, T., Sriprakash, K. S. and Radding, C. (1971). *Proc. natn. Acad. Sci. U.S.A.* **68**, 1639–1643.
Cerda Olmedo, E., Hanawalt, P. C. and Guerola, N. (1968). *J. molec. Biol.* **33**, 705–719.
Chandley, A. C. and Kofman-Alfaro, S. (1971). *Expl. Cell Res.* **69**, 45–48.
Chang, S., Flores, J. and Ponnamperuma, C. (1969). *Proc. natn. Acad. Sci. U.S.A.* **64**, 1011.
Clark, A. J. (1971). *Ann. Rev. Microbiol.* **25**, 437–464.
Clark, A. J. and Margulies, A. D. (1965). *Proc. natn. Acad. Sci. U.S.A.* **53**, 451–459.
Cleaver, J. E. (1969). *Int. J. Radiat. Biol.* **16**, 277–285.
Cleaver, J. E. (1970). *Int. J. Radiat. Biol.* **18**, 557–565.
Cleaver, J. E. and Thomas, G. H. (1969). *Biochem. biophys. Res. Commun.* **36**, 203–208.
Cleaver, J. E. and Trosko, J. E. (1969). *Int. J. Radiat. Biol.* **15**, 411.
Cleaver, J. E. and Trosko, J. E. (1970). *Photochem. Photobiol.* **11**, 547–550.
Cook, J. S. (1970). *In* "Photophysiology" (ed. Giese, A. C.), **5**, p. 191. Academic Press, New York and London.
Cook, J. S. and McGrath, J. R. (1967). *Proc. natn. Acad. Sci. U.S.A.* **58**, 1359.
Cook, J. S. and Regan, J. D. (1969). *Nature, Lond.* **223**, 1066–1067.
Cook, J. S. and Worthy, T. E. (1972). *Biochemistry* **11**, 388–393.
Cooper, P. K. and Hanawalt, P. C. (1972). *Proc. natn. Acad. Sci. U.S.A.* **69**, 1156–1160.
Cordone, L. and Radman, M. (1970). *Virology* **41**, 166–169.
Cox, B. S. and Parry, J. M. (1968). *Mutat. Res.* **6**, 37–55.
Cox, E. (1970). *J. molec. Biol.* **50**, 129–135.
Cox, E. C. and Yanofsky, C. (1967). *Proc. natn. Acad. Sci. U.S.A.* **58**, 1859–1902.
Crick, F. H. C. (1968). *J. molec. Biol.* **38**, 367–379.
Dean, C. and Pauling, C. (1970). *J. Bacteriol.* **102**, 588–589.
Deering, R. A. (1962). *Sci. Am.* **207**, 135.
Defais, M., Caillet-Fauquet, P., Radman, M. and Errera, M. (1971). *Virology* **43**, 495–503.
De Lucia, J. and Cairns, J. (1969). *Nature, Lond.* **224**, 1164–1166.
Donch, J., Green, M. H. L. and Greenberg, J. (1968). *J. Bacteriol.* **96**, 1704–1710.
Dove, W. F. (1968). *Ann. Rev. Genet.* **2**, 305–340.
Drake, J. W. (1969). *Ann. Rev. Genet.* **3**, 247–268.
Drake, J. W. (1970). *In* "The Molecular Basis of Mutation". Holden-Day, San Francisco, London, Cambridge, Amsterdam.
Drake, J. W., Allen, E. F. Forsberg, S. A., Preparata, R. M. and Greening, E. O. (1969). *Nature, Lond.* **221**, 1128–1132.
Echols, H. and Gingery, R. (1968). *J. molec. Biol.* **34**, 239–249.
Edenberg, H. and Hanawalt, P. (1972). *Biochim. biophys. Acta* **272**, 361–372.
Eigen, M. (1971a). *Naturwissenschaften* **58**, 465–522.
Eigen, M. (1971b). *Quart. Rev. Biophys.* **4**, 149–212.
Engel, W., Krone, W. and Wolf, U. (1967). *Mutat. Res.* **4**, 353–368.
Evans, N. A. and McLaren, A. D. (1969). *Photochem. Photobiol.* **9**, 243–246.
Fabre, F. (1970). *Mutat. Res.* **10**, 415.
Fabre, F. (1972a). *Mol. Gen. Genet.* **117**, 153–166.
Fabre, F. (1972b). *Photochem. Photobiol.* **15**, 367–373.

Ferris, J. P., Sanchez, R. A. and Orgel, L. E. (1968). *J. molec. Biol.* **33**, 693–704.
Fitch, W. M. (1971). *Nature New Biol.* **229**, 245–247.
Fornili, S. L., Cordone, L. and Radman, M. (1971). *Virology* **43**, 422–426.
Fortuin, J. J. H. (1971). *Mutat. res.* **11**, 265–277.
Fox, S. W. (1964). *Nature, Lond.* **201**, 336.
Fox, S. W. (1965). *In* "The Origine of Prebiological Systems". Academic Press, New York and London.
Freifelder, D. and Katz, G. (1971). *J. molec. Biol.* **57**, 351–354.
Friedberg, E. C. and King, J. J. (1971). *J. Bacteriol.* **106**, 500–507.
Friis, J., Flury, F. and Leupold, U. (1971). *Mutat. Res.* **11**, 373–390.
Fujiwara, Y. and Kondo, T. (1972). *Biochem. biophys. Res. Commun.* **47**, 557–564.
Garcia-Bellido, A. (1972). *Molec. Gen. Genet.* **115**, 54–72.
Gefter, M. L., Hirota, Y., Kornberg, T., Wechsler, J. A. and Barnoux, C. (1971). *Proc. natn. Acad. Sci. U.S.A.* **68**, 3150–3153.
Gellert, M. (1971). *In* "Procedures in Nucleic Acid Research" (ed. Cantoni, G. L. and Davies, D. R.), vol. 2, pp. 875–888. Harper and Row, New York.
Gellert, M. and Bullock, M. L. (1970). *Proc. natn. Acad. Sci. U.S.A.* **67**, 1580–1587.
Goebel, W. (1972). *Nature New Biol.* **237**, 67–70.
Goldmark, P. and Livinn, S. (1970). *Proc. natn. Acad. Sci. U.S.A.* **67**, 434–441.
Green, M. H. L., Donch, J. and Greenberg, J. (1969). *Photochem. Photobiol.* **10**, 397–405.
Gregoli, S., Taverna, C. and Bertinchamps, A. (1970). *Int. J. Radiat. Biol.* **18**, 577–585.
Griggs, H. G. and Bendor, M. A. (1972). *Photochem. Photobiol.* **15**, 517–526.
Gross, J. and Gross, M. (1969). *Nature, Lond.* **224**, 1166–1168.
Gross, J., Grunstein, J. and Witkin, E. (1971). *J. molec. Biol.* **58**, 631–634.
Haas, F. L. and Doudney, C. O. (1960). *Nature, Lond.* **185**, 637–638.
Harm, W. (1969a). *Mutat. Res.* **8**, 411–415.
Harm, W. (1969b). *Radiat. Res.* **39**, 517.
Harm, W., Harm, H. and Rupert, C. S. (1968). *Mutat. Res.* **6**, 371–385.
Haynes, R. H. (1964). *In* "Physical Processes in Radiation Biology" (eds Augenstein, L., Mason, R. and Rosenberg, B.), pp. 51–72. Academic Press, New York, and London.
Hedges, R. W. (1972). *Heredity* **28**, 39–48.
Helene, C. and Charlier, M. (1971). *Biochimie* **53**, 1175–1180.
Hidalgo Salviaterra, O. and McLaren, A. D. (1969). *Photochem. Photobiol.* **9**, 417–426.
Holliday, R. (1971). *Nature New Biol.* **232**, 233–236.
Holmquist, R., Jukes, T. H. and Pangburn, S. (1973). *J. molec. Biol.* **78**, 91–116.
Horii, I. and Suzuki, K. (1970). *Photochem. Photobiol.* **11**, 99–107.
Hotta, Y. and Stern, H. (1971). *J. molec. Biol.* **55**, 337–355.
Hotz, G. and Reuschl, H. (1967). *Mol. Gen. Genet.* **99**, 5–11.
Howard-Flanders, P. (1968). *Ann. Rev. Biochem.* **37**, 175–200.
Howard-Flanders, P. and Boyce, R. P. (1966). *Radiat. Res. Suppl.* **6**, 156–184.
Howard-Flanders, P. and Theriot, L. (1966). *Genetics* **53**, 1137–1150.
Howell, S. H. and Stern, H. (1971). *J. molec. Biol.* **55**, 357–378.
Huberman, J. A. and Riggs, A. D. (1968). *J. molec. Biol.* **32**, 327–341.
Hulett, H. R. (1969). *J. theor. Biol.* **24**, 56–72.
Hutchinson, F. and Hales, H. B. (1970). *J. molec. Biol.* **50**, 59–69.

Ibanez, J., Kimball, A. P. and Oro, J. (1971). *In* "Chemical Evolution and the Origin of Life" (eds Buvet, R. and Ponnamperuma, C.), pp. 171–179. Elsevier, North Holland, Amsterdam.

Ikenaya, M., Patrick, M. H. and Jagger, J. (1970). *Photochem. Photobiol.* **11**, 487–494.

Jacob, F. and Wollman, E.-L. (1955). *Annls Inst. Pasteur, Paris*, **88**, 724–749.

Jacobs, A., Bopp, A. and Hagen, U. (1972). *Int. J. Radiat. Biol.* **22**, 431–435.

Jagger, J. and Stafford, R. S. (1965). *Biophys. J.* **5**, 75–88.

Jagger, J., Stafford, R. S. and Snow, J. M. (1969). *Photochem. Photobiol.* **10**, 383–395.

Jagger, J., Tabeke, H. and Snow, J. M. (1970). *Photochem. Photobiol.* **12**, 185–196.

Jansen, G. J. O. (1970). *Mutat. Res.* **10**, 21–32.

Jansen, G. J. O. (1972). *Molec. Gen. Genet.* **116**, 47–50.

Jung, E. G. (1970). *Nature, Lond.* **228**, 361.

Kanazir, D. T. (1958). *Biochim. biophys. Acta* **30**, 20–23.

Kanazir, D. T. (1969). *Prog. Nucl. Acid. Res. Molec. Biol.* **9**, 117–222.

Kaplan, H. S. and Moses, L. E. (1964). *Science, Washington* **145**, 21–25.

Kato, R. (1967). *Hereditas* **58**, 221–247.

Kato, R. (1968). *Hereditas* **59**, 120–141.

Keiding, J. and Westergaard, O. (1971). *Expl Cell Res.* **64**, 317–322.

Kelly, R. B., Atkinson, M. R., Huberman, J. A. and Kornberg, A. (1969). *Nature, Lond.* **224**, 495–501.

Kelly, R. B., Cozzarelli, N. R., Deutscher, M. P., Lehman, I. R. and Kornberg, A. (1970). *J. biol. Chem.* **245**, 39–45.

Kimball, A. P. and Oro, J. (1971). *In* "Prebiotic and Biochemical Evolution". North Holland, Amsterdam.

Klimek, M. and Zemanova, L. (1971). *Neoplasma Bratislava* **18**, 87–97.

Kofman-Alfaro, S. and Chandley, A. C. (1971). *Expl. Cell Res.* **69**, 33–44.

Kondo, S., Ishikawa, H., Iwo, I. and Kato, T. (1970). *Genetics* **66**, 187–217.

Kushner, S., Kaplan, J. C., Ono, H. and Grossmann, L. (1971). *Biochemistry* **10**, 3325–3334.

Lamola, A. A. (1971). *Biochem. biophys. Res. Commun.* **43**, 893–898.

Latarjet, R. (1972). *Curr. Top. Radiat. Res.* **8**, 1–38.

Lawley, P. D. (1966). *Prog. Nucl. Acid. Res. Molec. Biol.* **5**, 89–131.

Lawrence, C. W. (1970). *Mutat. Res.* **10**, 557–566.

Lea, D. E. (1946). *In* "Action of Radiation on Living Cells". Cambridge University Press, London.

Lehmann, A. R. (1972). *J. molec. Biol.* **66**, 319–337.

Leigh, B. (1969). *Mutat. Res.* **8**, 101–109.

Lemontt, J. F. (1971). *Genetics* **68**, 21–33.

Lemontt, J. F. and Mortimer, R. K. (1970). *Genetics* **64**, S 37.

Lerman, L. S. (1963). *Proc. natn. Acad. Sci. U.S.A.* **49**, 94–102.

Lesk, A. M. (1969). *J. theor. Biol.* **22**, 537–540.

Lesk, A. M. (1970). *J. theor. Biol.* **27**, 171–173.

Leth Bak, A., Atkins, J. F. and Meyer, S. A. (1972). *Science, Washington* **175**, 1391–1393.

Ley, R. D. and Setlow, R. B. (1972). *Biophys. J.* **12**, 420–431.

Lieb, M. (1960). *Biochim. biophys. Acta* **37**, 155–157.

Lifschyts, E. and Falk, R. (1968). *Mutat. Res.* **6**, 235–244.

Lindahl, T. (1971a). *Eur. J. Biochem.* **18**, 407–414.
Lindahl, T. (1971b). *Eur. J. Biochem.* **18**, 415–421.
Lindahl, T. and Edelman, G. (1968). *Proc. natn. Acad. Sci. U.S.A.* **61**, 680–687.
Lindstrom, D. M. and Drake, J. W. (1970). *Proc. natn. Acad. Sci. U.S.A.* **65**, 617–624.
Loprieno, N. (1971). *Molec. Gen. Gent.* **110**, 348–354.
Luria, S. E. (1947). *Proc. natn. Acad. Sci. U.S.A.* **33**, 253–264.
Luria, S. E. (1952). *J. cell Comp. Physiol.* **39**, 119–123.
Luria, S. E. and Latarjet, R. (1947). *J. Bacteriol.* **53**, 149–163.
McCarthy, B. J. (1969). *In* "Handbook of Molecular Cytology" (ed. Lima de Faria, A.), pp. 4–20. North Holland, Amsterdam.
McReynolds, J. H., Furlong, N. B., Birrell, P. J., Kimball, A. P., Scheltgen, E. and Oro, J. (1971). *In* "Prebiotic and Biochemical Evolution" (eds Kimball, A. P. and Oro, J.), pp. 111–121. North Holland, Amsterdam.
Magni, G. E. (1969). *Proc. Int. Cong. Genet.* **3**, 247–259.
Magni, G. E., Von Borstel, R. C. and Sora, S. (1964). *Mutat. Res.* **1**, 227–230.
Malling, H. V. and de Serres, F. J. (1967). *Radiat. Res.* **31**, 637–638.
Manney, T. R. and Mortimer, R. K. (1964). *Science, Washington* **143**, 581–582.
Marin, G. and Prescott, D. M. (1964). *J. cell Biol.* **21**, 159–167.
Masamune, Y. and Richardson, C. C. (1968). *Proc. natn. Acad. Sci. U.S.A.* **61**, 1328–1335.
Meistrich, M. L. and Drake, J. W. (1972). *J. molec. Biol.* **66**, 107–114.
Mennigmann, H. D. and Szybalski, W. (1962). *Biochem. biophys. Res. Commun.* **9**, 398–404.
Meselson, M. S. (1967). *In* "Heritage from Mendel" (ed. Brink, R. A.), pp. 81–103. The University of Wisconsin Press.
Miller, S. L. (1955). *J. Am. chem. Soc.* **77**, 2351–2360.
Miller, S. L. and Urey, H. C. (1959). *Science, Washington* **130**, 245–251.
Mills, D. R., Kramer, F. R. and Spiegelman, S. (1973). *Science, N.Y.* **180**, 916–927.
Minato, S. and Werbin, H. (1971). *Biochemistry* **10**, 4503–4508.
Miura, A. and Tomizawa, J. (1968). *Molec. Gen. Genet.* **103**, 1–10.
Mortimer, R. K. (1958). *Radiat. Res.* **9**, 312–326.
Mourant, A. E. (1971). *Nature, Lond.* **231**, 466.
Muller, H. J. (1967). *In* "Heritage from Mendel" (ed. Brink, R. A.), p. 419. University of Wisconsin Press, Madison.
Murphy, T. M. and Gordon, M. P. (1971a). *Photochem. Photobiol.* **13**, 45–55.
Murphy, T. M. and Gordon, M. P. (1971b). *Photochem. Photobiol.* **14**, 721.
Nasim, A. (1968). *Genetics* **59**, 327–333.
Neuman, M. W., Neuman, W. T. and Lane, K. (1970). *Curr. Mod. Biol.* **3**, 253–259, 277–283.
Ohno, S. (1970). *In* "Evolution by Gene Duplication". Springer Verlag, Berlin, Heidelberg, New York.
Ohshikama, S. and Sekigushi, M. (1972). *Biochem. biophys. Res. Commun.* **47**, 1126–1132.
Ohta, T. and Kimura, M. (1971). *Nature, Lond.* **233**, 118–119.
Okada, Y., Streisinger, G., Owen, J. E., Newton, J., Tsugita, A. and Ihoye, M. (1972). *Nature, Lond.* **236**, 338–341.
Ono, J. and Shimazu, Y. (1966). *Virology* **29**, 295–302.
Orgel, L. E. (1968). *J. molec. Biol.* **38**, 381–393.

Orgel, L. E. and Sulston, J. E. (1971). *In* "Prebiotic and Biochemical Evolution" (eds Kimball, A. P. and Oro, J.), pp. 89–94. Elsevier, North Holland, Amsterdam.

Oro, J. (1965). *In* "The Origins of Prebiological Systems and their Molecular Martices" (ed. Fox, S. W.). Academic Press, New York and London.

Paecht-Horowitz, M., Berger, J. and Katchalsky, A. (1970). *Nature, Lond.* **228**, 636–639.

Painter, R. B. (1970). *Curr. Top. Radiat. Res.* **7**, 45–70.

Parry, J. M. (1971). *Molec. Gen. Genet.* **111**, 51–60.

Paterson, M. C. and Roozen, K. J. (1972). *J. Bacteriol.* **110**, 71–80.

Pathak, M. A. (1972). *VIth Int. Cong. Photobiol., Bochum* 21–25/8/72.

Pauling, C. and Hamm, L. (1968). *Proc. natn. Acad. Sci. U.S.A.* **60**, 1495–1502.

Petes, T. D. and Fangman, W. L. (1972). *Proc. natn. Acad. Sci. U.S.A.* **69**, 1188–1191.

Phillips, J. H. and Brown, D. M. (1967). *Prog. Nucl. Acid. Res. Molec. Biol.* **7**, 349–368.

Pietrzykowska, I. (1973). *Mutat. Res.* **19**, 1–9.

Ponnamperuma, C. (1965). *In* "The Origins of Prebiological Systems and Their Molecular Matrices" (ed. Fox, S. W.), pp. 221–242. Academic Press, New York and London.

Ponnamperuma, C., Sagan, C. and Mariner, R. (1963). *Nature, Lond.* **199**, 222–226.

Pörschke, D. and Eigen, M. (1971). *J. molec. Biol.* **62**, 361–381.

Proust, J. (1967). *Mutat. Res.* **4**, 837–863.

Prudhommeau, C. and Proust, J. (1969). *Mutat. Res.* **8**, 317–326.

Pullman, B. and Pullman, A. (1962). *Nature, Lond.* **196**, 1137–1142.

Rabinowitz, J., Chang, S. and Ponnamperuma, C. (1971). *In* "Prebiotic and Biochemical Evolution" (eds Kimball, A. P. and Oro, J.), pp. 70–77. Elsevier, North Holland, Amsterdam.

Radman, M. (1971). *Nature New Biol.* **230**, 277–278.

Radman, M. and Errera, M. (1970). *Mutat. Res.* **9**, 553–560.

Radman, M., Cordone, L., Krsmanovic-Simic, D. and Errera, M. (1970). *J. molec. Biol.* **49**, 203–212.

Rasmussen, R. E. and Painter, R. B. (1966). *J. cell Biol.* **29**, 11–19.

Rauth, A. M. (1967). *Radiat. Res.* **31**, 121–138.

Regan, J. D., Trosko, J. E. and Carrier, W. L. (1968). *Biophys. J.* **8**, 319–325.

Renz, M., Lohromann, R. and Orgel, L. E. (1971). *Biochim. biophys. Acta* **240**, 463–471.

Resnick, M. A. (1969). *Genetics*, **62**, 519–531.

Resnick, M. A. (1970). *Nature, Lond.* **226**, 337.

Richardson, C. C. (1969). *A. Rev. Biochem.* **38**, 795–840.

Roman, H. and Jacob, F. (1958). *Cold Spring Harb. Symp. quant. Biol.* **23**, 155–160.

Rommelaere, J. and Errera, M. (1972). *Int. J. Radiat. Biol.* **22**, 285–291.

Rommelaere, J., Susskind, M. and Errera, M. (1973). *Chromosoma* **41**, 243–257.

Rupp, W. D. and Howard-Flanders, P. (1968). *J. molec. Biol.* **31**, 291–304.

Rupp, W. D., Wilde III, C. E., Reno, D. L. and Howard-Flanders, P. (1971). *J. molec. Biol.* **61**, 25–44.

Sagan, C. (1965). *In* "The Origins of Prebiological Systems and their Molecular Matrices" (ed. Fox, S. W.), pp. 207–219. Academic Press, New York and London.

Sagan, C. and Khare, B. N. (1971). *Science, Washington* **173**, 417–420.

Saito, N. and Werbin, H. (1969). *Photochem. Photobiol.* **9**, 389–393.

Sanchez, R. A., Ferris, J. P. and Orgel, L. E. (1966). *Science, Washington* **154**, 784–785.

Saucier, J. M., Festy, B. and Le Pecq, J. B. (1971). *Biochimie* **53**, 973–980.

Savić, D. J. and Kanazir, D. T. (1972). *Molec. Gen. Genet.* **118**, 45–50.

Sawada, S. and Okada, S. (1970). *Radiat. Res.* **41**, 145–162.

Saxinger, C. and Ponnamperuma, C. (1971). *J. molec. Evolution* **1**, 63–73.

Schneider-Bernloehr, H., Lohrmann, R., Sulston, J., Weimann, B. J., Orgel, L. E. and Miller, H. T. (1968). *J. molec. Biol.* **37**, 151–155.

Schwartz, A. and Ponnamperuma, C. (1971). *In* "Prebiotic and Biochemical Evolution" (eds Kimball, A. P. and Oro, J.), pp. 78–82. Elsevier, North Holland, Amsterdam.

Sesnowitz-Horn, S. and Adelberg, E. A. (1968). *Cold Spring Harb. Symp. quant. Biol.* **33**, 393–402.

Setlow, J. K. (1966). *Radiat. Res. Suppl.* **6**, 141–155.

Setlow, J. K. and Bollum, F. J. (1968). *Biochim. biophys. Acta* **157**, 233–237.

Setlow, P. and Kornberg, A. (1972). *J. biol. Chem.* **247**, 232–240.

Setlow, P., Brutlag, D. and Kornberg, A. (1972). *J. biol. Chem.* **247**, 224–231.

Setlow, R. B. (1968). *Prog. Nucl. Acid. Res. Molec. Biol.* **8**, 257–295.

Setlow, R. B. and Carrier, W. L. (1966). *J. molec. Biol.* **17**, 237–254.

Setlow, R. B. and Setlow, J. K. (1962). *Proc. natn. Acad. Sci. U.S.A.* **48**, 1250–1257.

Shafronskaya, N. N., Trifonov, E. N., Lazurkin, Y. S. and Frank-Kamenetsky, M. D. (1973). *Nature New Biol.* **241**, 58–60.

Shanfield, B. and Kafer, E. (1971). *Genetics* **67**, 209–219.

Sharp, J. D. and Freifelder, D. (1970). *IVth Int. Cong. Radiat. Res. Evian*, 771.

Signer, E. R. and Weil, J. (1968). *J. molec. Biol.* **34**, 261–271.

Singer, C. E. and Ames, B. N. (1970). *Science, Washington* **170**, 822–826.

Slizynska, Y. (1969). *Mutat. Res.* **8**, 165–175.

Smirnov, G. B., Favoiskaya, Y. U. N. and Stravonskaya, A. V. (1971). *Molec. Gen. Genet.* **111**, 357–367.

Smith, K. C. (1971). *In* "Photophysiology" (ed. A. C. Giese) Academic Press, New York and London.

Smith, K. C. and Hanawalt, P. C. (1969). *In* "Molecular Photobiology". Academic Press, New York and London.

Smith, K. C. and Meun, D. H. C. (1970). *J. molec. Biol.* **51**, 459–472.

Spadari, S., Ciarrochi, G. and Falaschi, A. (1971). *Eur. J. Biochem.* **22**, 75–78.

Spiegelman, S. (1971). *Quart. Rev. Biophys.* **4**, 213–253.

Spiegler, P. and Norman, A. (1969). *Radiat. Res.* **39**, 400–412.

Springate, C. F. and Loeb, L. A. (1973). *Proc. natn. Acad. Sci. U.S.A.* **70**, 245–249.

Steinman, G., Smith, A. E. and Silver, J. J. (1968). *Science, Washington* **159**, 1108–1109.

Stephen Sherwood, E., Oro, J. and Kimball, A. P. (1971). *Science, Washington* **173**, 446–447.

Streisinger, G., Okada, Y., Emrich, J., Newton, J., Tsugita, A., Terzaghi, E. and Inouye, M. (1966). *Cold Spring Harb. Symp. quant. Biol.* **31**, 77–84.

Sugimura, T., Otake, H., Matsushima, T. (1968). *Nature, Lond.* **218**, 392.

Sugino, A. and Okazaki, R. (1972). *J. molec. Biol.* **64**, 61–85.

Sulston, J., Lohrmann, K., Orgel, L. E., Schneider-Bernloehr, H., Weiman, B. J. and Milles, H. T. (1969). *J. molec. Biol.* **40**, 227.

Sutherland, B. M. and Sutherland, J. C. (1969). *Biophys. J.* **9**, 292–302, 1045–1055, 1329–1336.
Tapiere, C. M. and Nagyvary, J. (1971). *Nature, Lond.* **231**, 42–43.
Terzi, M. (1961). *Nature, Lond.* **191**, 461–462.
Treffers, H. P., Spinelli, V. and Belser, N. O. (1954). *Proc. natn. Acad. Sci. U.S.A.* **40**, 1064–1071.
Trosko, J. E. and Isoun, M. (1970). *Int. J. Radiat. Biol.* **18,** 271–275.
Trosko, J. E., Chu, E. H. Y. and Carrier, W. L. (1965). *Radiat. Res.* **24**, 667–672.
Tsukuda, K. and Niski, A. (1971). *Biochem. J.* **70**, 541–542.
Unrau, P., Wheateroft, R. and Cox, B. S. (1971). *Molec. Gen. Genet.* **113**, 359–362.
Von Borstel, R. C., Graham, D. E., La Brot, K. J. and Resnick, M. A. (1968). *Genetics* **60**, 233.
Watson, W. A. (1972). *Mutat. Res.* **14**, 299–307.
Weigle, J. (1953). *Proc. natn. Acad. Sci. U.S.A.* **39**, 628–636.
Weiss, B. and Richardson, C. C. (1967). *Proc. natn. Acad. Sci. U.S.A.* **57**, 1021–1028.
Werbin, H. and Rupert, C. S. (1967). *Photochem. Photobiol.* **7**, 225–230.
Werbin, H., Valentine, R. C., Hidalgo Salviaterra, O. and McLaren, A. D. (1968). *Photochem. Photobiol.* **7**, 253–261.
Westergaard, O. and Pearlman, R. E. (1969). *Expl Cell Res.* **54**, 309–313.
Witkin, E. M. (1967). *Brookhaven Symp. Biol.* **20**, 17–53.
Witkin, E. M. (1969a). *A. Rev. Microbiol.* **23**, 487–514.
Witkin, E. M. (1969b). *Mutat. Res.* **8**, 9–14.
Witkin, E. M. (1971). *Nature New Biol.* **229**, 81–82.
Woese, C. R., Dugre, D. H., Dugre, S. A., Kondo, M. and Saxinger, W. C. (1966). *Cold Spring Harb. Symp. quant. Biol.* **31**, 723–726.
Wood, S. and Käfer, E. (1969). *Genetics* **62**, 507–518.
Woodcock, E. and Grigg, G. W. (1972). *Nature New Biol.* **237**, 76–79.
Wright, M., Buttin, G. and Hurwitz, J. (1971). *J. biol. Chem.* **246**, 6543–6555.
Wulff, D. L. and Rupert, C. S. (1962). *Biochem. biophys. Res. Commun.* **7**, 237–240.
Yanofsky, C., Ito, J. and Horn, V. (1966). *Cold Spring Harb. Symp. quant. Biol.* **31**, 151–162.

Author Index

Numbers in *italics* refer to the pages on which references are listed at the end of each chapter

I

J

K

Subject Index

E